Efficiency and the Social Services

Efficiency and the Social Services

Robert Pruger
Editor

The Haworth Press
New York • London

Efficiency and the Social Services has also been published as *Administration in Social Work*, Volume 15, Numbers 1/2, 1991.

The Haworth Press, Inc. 10 Alice Street, Binghamton, NY 13904-1580
EUROSPAN/Haworth, 3 Henrietta Street, London WC2E 8LU England

Library of Congress Cataloging-in-Publication-Data

Efficiency and the social services / Robert Pruger, editor.
p. cm.
"Efficiency and the social services has also been published as Administration in social work, volume 15, numbers 1/2 1991."
ISBN 1-56024-113-6 (alk. paper)
1. Social work administration. 2. Social service – Cost effectiveness. I. Pruger, Robert.
HV40.E4 1990
361'.0068 – dc20

91-2035
CIP

Efficiency and the Social Services

CONTENTS

∞ ALL HAWORTH BOOKS & JOURNALS ARE PRINTED ON CERTIFIED ACID-FREE PAPER

ABOUT THE EDITOR

Robert Pruger, DSW, is Professor in the School of Social Welfare, the University of California, Berkeley, where he currently teaches a class called "Efficiency in Welfare Administration." He has spent more than 15 years working with economist Leonard Miller, trying to clarify how microeconomic theory and its central concept, efficiency, could illuminate the social services. The articles "The Two Activities of Social Services: Maintenance and People Changing" (1989), "The Division of Labor in a Perfect Maintenance Agency" (1990), and "Efficiency in Long Term Care: The Case of Dual Objectives" (1990), are his three most recent publications out of the 10 articles he has written on the topic. Dr. Pruger earned his MSW from the University of Pennsylvania and his DSW from the University of California, Berkeley.

Efficiency and the Social Services

Introduction

Robert Pruger, DSW

My most important duty as guest editor (or so it seemed) was to nag authors about deadlines. There was no fun in it but, thankfully, the contributors to this volume never turned crabby. Even if the opposite were true, however, it would have been worth it. The nagging helped to produce a volume on a topic – efficiency applied to the social sciences – for which almost no serious literature exists.

All of the following can be said about efficiency. It is a measure of how well *any* purpose is achieved. This universal relevance and the complexities involved explain why an entire academic discipline and body of theory are devoted to explicating it. The emerging global economy is forcing greater attention to it among even very practical persons, as are tight budgets and rising deficits at all governmental levels. The Nobel Prize is annually awarded to those whose work promises to lead to more of it; and largely because it had too little of it, the communist empire is disintegrating. The social services bureaucracy, too, is a "command economy" with substantial efficiency problems. Yet, very few who operate, regulate, or study these agencies have a good grasp of the concept or act as if it is important.

The inattentiveness begins at the beginning: social work students may not even hear the idea mentioned during their years of graduate education, no less be helped to achieve a meaningful level of comprehension. Many welfare administration texts do not discuss it at all, and the treatment in most of the others aims no higher than to distinguish it from effectiveness (and even that is done in an *ad hoc*, inconsequential way). What could be the invitation to a career-long

---

Dr. Pruger is Professor, School of Social Welfare, University of California at Berkeley, Berkeley, CA 94720.

intellectual and operational adventure is reduced to a virtually pointless classroom memorization exercise. As might be expected, efficiency rarely is the subject of a doctoral dissertation in social welfare. In the general literature of the field, the term is most frequently placed in a negative context, usually in the form of complaint against cost-cutting efforts—so called "efficiency drives"—that have little or nothing to do with efficiency.

The efficiency criterion can be applied to the social service agency over a wide range of its functioning. At the simplest level, it draws attention to waste that is so obvious as to be detectable by the eye alone, or at least requires no more knowledge than that acquired through routine experience on the job: forms that have no purpose; files too disorganized to yield needed information in a timely manner; regulations untouched by the reality of the circumstance being regulated. Deeper levels of application do not flow from such self-evident chaos. An illustrative concern here is the degree to which social workers are treating each client with the optimum amount and kinds of each service the agency provides. Except in the grossest instances, inefficiencies over this range cannot be directly observed; revealing them requires special skills and analytic techniques. The extra effort this involves will, in turn, be made only where there is a prior, more informed commitment to increase service efficiency.

At a still deeper level there is the matter of the design of the entire delivery structure. The key issue here is how well the system produces and distributes information to its discretionary decision makers—social workers, supervisors, and managers—and the need to make better resource allocations over time. This is the most exciting issue that exists at the frontier of the search for knowledge to administer better the helping enterprise. Unfortunately, it also goes well beyond what most think is encompassed by the efficiency concept.

Efficiency may never be treated, as it should be, as a fundamental standard in the social services. This volume can only be a small step in the right direction. Its mere existence, at least implicitly, connotes that there is a subject here worthy of attention. A close reading of its contents should add the additional sense that when efficiency is applied to the social services there are theoretical questions and technological challenges that are as fascinating and demanding as any academics and practitioners are likely to come upon.

The collection is unavoidably unbalanced in its coverage of the terrain. This is because those who prepared manuscripts for this volume had to proceed without benefit of being anchored in an ongoing community of scholarly interest in efficiency in the social services. This long inattention results in no agreement about what the major issues are and no readily recognized group of experts to turn to for authoritative analyses of them. While almost all authors are either on social work faculties or from organizations concerned with the social services, the majority do not hold social work degrees. When multidisciplinary attentions are directed to a topic for which there is little shared understanding of what most needs to be addressed, strong editorial guidance of authors is not warranted. Perhaps the future will see more systematic attention paid to efficiency by welfare academics and operatives. If so, editors, with much greater comfort and success than could be enjoyed this time around, will be able to nag writers into line. I can hardly wait.

PART 1: INTRODUCING THE CONCEPT

Efficiency and the Social Services: Part A

Robert Pruger, DSW
Leonard Miller, PhD

Efficiency is both a conceptual tool and a vision of perfection. In the social welfare enterprise; however, the tool is not used and the vision is neither understood nor valued. Indeed, negative references are common. Gilbert (1983), for example, observes that "the idea of 'efficient benevolence' sounds faintly contradictory" (p. 28). Others are less restrained. Rees and Wallace (1984) "criticize (social service) evaluations that emphasize . . . efficiency" (p. 309). And Fabricant (1985) warns that the "cost of greater efficiency" is the loss of "the quality, content and outcome of encounter between (social) workers and citizens" (p. 392).

Of course, not every mention is darkly colored. Nevertheless, one must turn to outsiders for an informed, unambiguously positive characterization. A subsequent winner of the Nobel Prize, for example, described the efficiency principle as nothing less than the ultimate organizational desideratum. It was, he said, a definition of:

Dr. Pruger and Dr. Miller are Professors, School of Social Welfare, University of California at Berkeley, Berkeley, CA 94720.

> . . . what is meant by 'good' or 'correct' administrative behavior. It does not tell *how* accomplishments are to be maximized, but merely states that this maximization is the aim of administrative activity. (Simon, 1987) (emphasis in original)

An authoritative guide to the English language (Webster, 1961) also gives the term an optimistic coloration. Efficiency, it says, is synonymous with "efficaciousness, effectiveness, efficacy, power, proficiency, skill, ability, capability, competence." Its opposites are "inefficiency, incompetence, impotence, inability."

Though efficiency never has been treated as a first order concept in the social services, it should be. No other idea on the table or on the horizon cuts so directly to the core of the enterprise. No other has the same potential to revitalize the field as it moves into and through the 21st century. For that to occur, however, efficiency will have to be taken more seriously than it has been in the past. That begins with a proper understanding of the term.

WHAT EFFICIENCY IS–AND WHAT IT IS NOT

Efficiency, a concept, is important because scarcity, a condition of the real world, exists. The concept is needed to help us manage the condition. In social welfare, scarcity is most easily recognized as not having enough time, money, staff or other resources to accomplish organizational or service objectives. In this form, it probably is experienced by every worker every day at every level of the enterprise. There are, however, two different kinds of scarcity, and a different treatment is appropriate to each. The difference between them yields the foundation of the argument being presented here.

Political Scarcity

This refers to a felt lack of resources attributable to inadequate allocations made by some higher authority, often because another activity was funded too generously. For example, one often hears that Congress has provided too little for social welfare programs because it provided too much for defense. Regarding the social welfare budget, some argue that too little is spent on children because too much is allocated to the elderly. And within all organizations, bureaus regularly compete to have their budget share enlarged even though it necessarily means that other bureaus will receive lesser shares.

Three key points can be made about political scarcity. First, the resource deficiency is treated as being temporary and soluable. Thus, Hasenfeld (1984) is able to write, "If one assumes that the economy of scarcity will persist . . ." (p. 525). And Algen (1980) can say, "The present conservative swing, with its rhetoric on the merits of scarcity, is a pendulum that I suggest we, as mental health workers, should view with especial alarm" (p. 392).

Second, the solution is to get more resources. As Niskanen (1971) points out, the bureaucrat seeks to maximize the budget, rather than the value of what one produces. Third, larger resource shares are obtained by pressing higher authority to provide them. Toward this end, evidence of rational performance is never sufficient, and often it is not even necessary. As one state official, who understood his major responsibility to be to protect the budget of the program he led, insisted, when we discussed these matters with him, "Efficiency and effectiveness are luxuries I just don't have time for" (Pruger, 1987, p. 40). At the time, we thought this was outrageous, but essentially he was right. Political scarcity is appropriately treated as a political problem best met through political means:

> Kogut has argued that, rather than accommodate to disinvestment in social welfare, coalitions should be formed to alter the political context . . . . The extent to which resources are scarce, he continues, is affected by policy . . . . At present, it is probably wise to reject economic analysis and pursue political action to alter the political context. When the nation is restored to the pursuit of both efficiency and improved social welfare, rational decision making may be tried again. (Buxbaum, 1985, p. 76)

Economic Scarcity

Even if politically skilled operatives win larger resource shares – indeed even if they get much more than they sought – resources will still be felt to be scarce. This is because no matter how much is spent there will always be one more thing worthy of accomplishment: one more client to serve or serve better; one more program to start or enlarge; one more problem to solve. This kind of scarcity—economic scarcity—is permanent and insolvable. It exists no matter what the size of the budget; for, as the resource pool increases, so does the set of objectives being pursued.

> Just as the kinds of goals sought must affect the means for achieving them, so the available resources (the ubiquitous money, knowledge and power) must shape the objectives sought. A billion-dollar program is not just larger than a million-dollar one; it may try to do different things in different ways. (Wildavsky, 1976, p. 50)

But even if objectives somehow remained fixed and the budget corresponded to what was needed to achieve them completely, economic scarcity would remain. Resource managers would still have a dual obligation to discharge: (a) find a way to achieve those same results at a lower cost, and/or (b) improve the quality or quantity of those results (e.g., provide clients with even more or better services). The latter increases the value of the undertaking; the former releases resources to achieve other desired ends currently less well provided for.

Social Work and Economic Scarcity

Social workers respond in a much livelier way to political scarcity than to economic scarcity. There is a greater sensitivity to its existence, and a stronger emotional, ideological and intellectual reaction to evidence of success and failure. There is also a greater readiness to learn how to deal with it. Graduate social work students, for example, are far more likely to enroll in courses dealing with how to increase resources (e.g., advocacy, proposal writing) than with those concerned with how to use a fixed amount.

Social workers' preferences do not serve them well in this matter, for they really have no choice here. As irony would have it, they will spend almost their entire careers allocating existing budgets; conversely, they will spend very little time getting those resource shares augmented. As it might be put, budgets, enlarged or shrunken, arrive once a year. The rest of the time is necessarily spend trying to get the most out of what arrived.

Consider the most ordinary, and therefore quite important, circumstance. Every day every caseworker must allocate (or have allocated for him/her) a time budget of eight hours – so much for clients (and within that so much for Client A and so much for Client B), and administrative work (and within that so much for the committee meeting, so much for record keeping, so much for trying to get the supervisor to change the way the worker's time has been assigned). Except in relatively insignificant, only temporarily available ways (e.g., take a short lunch "hour," skip the coffee break), there is no possibility of increasing this budget.

The situation is exactly the same even when getting resources increased is the whole point of the game (Lohman, 1984). This is because the attainment of this objective, like the attainment of all objectives, comes up against the most relentless scarcity of all – time. Thus, to accomplish his/her political end of "resource mobilization," the resource mobilizer, like the caseworker, must learn to evaluate the relative time costs and likely gains involved in each possible "intervention." For example, to raise the needed funds: launch a mass mailing appeal; arrange a dinner for selected large contributors; or submit a proposal to a local foundation. To get a desired program approved: testify at a legislative hearing; organize others in the community to do so; or generate media interest in the problem the program will address. To increase attendance at some rally: phone residents who will contact their neighbors; distribute leaflets at every busy street corner; or cover the neighborhood with posters announcing the event. Every one of these possibilities, and many others that would occur to the skillful worker, would help achieve the resource mobilization objective. But there will never be enough time to try them all.

All of this is also true at the agency level. Indeed, the complexities here are quite daunting: How should I allocate the time of an entire staff? How should I assess the productivity of all the interventions employed and those that might be tried? How should I establish the relative valuation of the different possible client groups served and service objectives adopted? And, most of all, once having decided all these matters, how should I adjust those decisions continuously over time to account for ever changing conditions?

In spite of all this, one subset of the profession has an especially difficult time taking economic scarcity seriously. Their ideological leanings get in the way. It is not so much that they deny its existence. Rather, they respond to any call to attend to economic scarcity as a denial of political scarcity. And this they seem to experience as a betrayal of sorts. Typically, they attempt to convert any discussion of the former into a discussion of the latter:

> The article by Edward Newman and Jerry Turem titled 'The Crisis of Accountability' raised again an issue that in our opinion has assumed too much prominence in recent social work journals: to wit, that social workers have failed to demonstrate a responsible attitude in evaluating the relationships of cause and effect in their intervention. We view this as a dysfunctional trend in social work literature, one that ignores the broader spectrum of the society in which social workers provide services. Ignoring the political context leads to support of the political *status quo*. . . . We do not see how this article assists social workers in dealing more adequately with issues of redistribution. (Henderson & Reid, 1974, pp. 387, 507)

Politics can increase resource shares from less to more, but it cannot help once it is recognized, as inevitably it must be, that more is always less than is needed. In brief, the problem of economic scarcity requires not political action but rational use of the available resources. Another term for that is efficiency.

What Efficiency Is

The search for efficiency is the search for a better relationship between means and ends. When more is accomplished with the same amount of resources; or, when the same result is accomplished with fewer resources, that better relationship has been achieved. The former can be thought of as efficiency through maximization; the latter as efficiency through minimization. The concept of efficiency includes both approaches to improvement.

The important thing to understand is that a statement about efficiency is a statement about input AND outcome, means AND ends, costs AND benefits, even if as a conversational convenience only one is mentioned. Output maximization bespeaks greater efficiency *only if costs have not risen proportionately*. Exactly analogously, lower costs do not indicate greater efficiency *unless output has not correspondingly declined*. This point must be made strongly because so many speak and write as if they did not understand it. This is the most common error made in the use of the term. In addition, it probably accounts more than any other for the indifference and negative reaction the concept evokes in the field.

What Efficiency Is Not

Consider the following excerpts, all of which are taken from the same book by a leading author in the field (Gilbert, 1983). The message of the first set is that, for a variety of weighty reasons, efficiency cannot yet widely serve as a criterion for evaluating social service programs. The message of the second set suggests that the opposite is true:

EFFICIENCY CANNOT BE USED

1. "Social welfare programs often serve objectives that are impalpable and multiple. The objectives frequently defy precise measurement. Many social welfare programs have been studied but relatively few unambiguous measures of effectiveness have been produced." (pp. 11-12)
2. "Given the multiple and often vague character of social welfare goals. . ."(p. 14)
3. "Efficiency can be a deceptive criterion in the social market, where . . . measures of effectiveness are complex and difficult to obtain. Surrogate outcomes, such as number of clients served, often displace assessment of substantive benefits, creating an arbitrary base against which to measure efficiency." (p. 26)

EFFICIENCY CAN BE USED

1. ". . . public funding agencies have put increasing pressure on service providers to improve the efficiency of social welfare programs. . .The emphasis on efficiency has continued as reflected in the Office of Human Development Service research priorities." (p. 28)
2. "It is, of course, quite reasonable to require agencies in the social market to operate as efficiently as possible." (p. 28)
3. ". . . nonservice approaches travel along fairly conventional routes. In the main they aim at creating highly efficient delivery systems by stretching available resources and supplementing services with low cost alternatives." (p. 30)

4. ". . . which brings us back to the problem of grasping for easily measured standards of efficiency in the subtler and more complex measures of program effectiveness—a situation in which decisionmakers learn the price of everything and the value of nothing" (p. 31)

How can what cannot be done be done? The answer is, it cannot. The contradiction, however, is apparent rather than real. It is explained by the fact that different meanings of efficiency (discussed below) are employed. The term is used in so many different ways and with so many different colorations that even a seasoned welfare scholar can get tangled in them.

Our definition, presented earlier, treats efficiency as a tool to aid in the management of scarcity. This is the meaning used in economics. It focuses attention on that discipline's central intellectual problem, the administration of scarce resources. It is equally appropriate to those responsible for the administration of scarce social service resources.

Next we examine several stated or implied meanings that illustrate what efficiency is *not*. They are drawn from the social services literature. Collectively, they represent one reason the concept is underused and identify how it is misused in the field.

1. Efficiency does not mean cost cutting or cost containment. Plans and aspirations, as expressed in budgets, ultimately must be paid for by available revenue. When the latter falls short, pressure to cut or control costs ineluctably builds. No other pressure so readily attracts and holds the attention of administrators, legislators and others with serious governance responsibilities.

Fiscal retrenchment is never happy work. Particularly over the short term, it can be the only responsible or even possible action (Taubman, 1985). Though necessary, it may also be thankless. On the other hand, creative managers may sometimes find something welcome in it (Knighton & Heidelman, 1984). Whatever it is, however, the relevant point here is that it is not efficiency; it is simply cost cutting.

As stated earlier, efficiency refers to a relationship between costs and benefits. Thus, if cost reduction is accompanied by disproportionately large reductions in the quality or quantity of the results, *in*efficiency has increased. Neugeboren (1985) expresses a view widely held in the field, but it is erroneous to say, "Efficiency is defined as stress on cost reduction and conservation of resources" (p. 202). Similarly, Gilbert (1983) is correct when he acknowledges that both costs and benefits are involved, but incorrect when he implies that a focus on only one expresses a concern for efficiency:

> Given the difficulty of evaluating exactly what many social welfare programs achieve, funding sources naturally turn to comparisons of how much it costs to run a program and support the most efficient operations. (p. 28)

Judge and Knapp (1985) have it entirely right. They recognize the need to find lower cost ways to operate, but they also tie that accomplishment to the continued achievement of service goals:

> The persistent fiscal pressure facing modern welfare states places a considerable premium on the search for efficiency gains. One of the most urgent questions in domestic public policy debate is whether the organization and production of social services can be modified in ways which promote goals at lower costs. (p. 131)

In a field that is philosophically liberal and humanitarian, equating efficiency with low or reduced budgets is not merely an error; it also prejudices the case. It improperly converts a natural upset about inadequate budgets into a negative reaction to the efficiency concept. Cost cutting is cost cutting. Those who implement or write about it should call it that. Terms like "efficiency drive" ought to be reserved for the real thing.

2. *Efficiency does not mean businesslike*. Social welfare management practices do not easily keep up-to-date with the increasingly complex tasks assigned. Thus, social service organizations are unusually vulnerable to the charge that they need to sharpen the way they operate. In a word, that they should be more "businesslike."

The term does not have a specific meaning. If there is any emphasis at all it is on the input side. Thus, to be businesslike is to do things like computerize operations; tighten management controls; discover and eliminate needless paperwork and redundant programs and staff; spruce up lines of communication; and the like.

Sometimes it means putting businessmen in positions of influence. In such cases social services staff are likely to complain that the newcomers have no understanding of social service objectives. And the latter are likely to insist with equal irritation that the former know nothing about costs. Neither, however, may care much about efficiency, which involves the simultaneous consideration of both objectives and costs.

The relevant point, then, is almost exactly the same as was made about cost cutting. Being businesslike can lead to greater efficiency, *or greater inefficiency* (e.g., poorly designed computerization efforts that regularly make everything worse) (Dery, 1981). Or, it can result in a marginally better or worse set of administrative practices that makes no discernible impact on efficiency. In any case, the terms have no necessary relationship with each other, least of all one of equivalent meaning.

Equating the adoption of business methods with efficiency also adds to the negative burden the concept must carry. Good (1987), for example, finds that brief therapy, "the lean – and sometimes mean – treatment" is entirely compatible with the values of "the business world's bottom-line of time management, cost efficiency, taking control, getting ahead, and staying on top" (p. 9). Hansenfeld's (1984) warning is more comprehensive:

> Cutback management forces human service organizations to consider strategies that stress efficiency and cost-effectiveness and to adopt the rhetoric of the private business corporation, although such strategies and styles may be in direct conflict with the mandate, domain, and service technology of the service organization. (p. 527)

3. Efficiency does not mean the achievement of irrational standards. Higher authority's most common way to solve problems in programs run by lower authority is the promulgation of tighter regulations. These rules often express the former's taking charge of the problem. They also announce and celebrate the authority's tough, no nonsense, results-oriented approach to management.

Consider, for example, a well-known attempt at standard setting by the federal government:

> This (drive for efficiency) was set in motion during the Nixon administration, when old-line professional social welfare leaders in the Department of Health Education and Welfare (HEW) were replaced by management-oriented officials, whose first act was a strenuous campaign to reduce error rates in public assistance. (Gilbert, 1983, p. 27)

The new regulations, backed by sanctions, announced lower benefit payment error rate standards. Though rationalized in terms of greater efficiency, the effort's design did not reflect even a minimum understanding of that term. A witness at a legislative oversight hearing made that quite clear by reminding everyone of an elementary lesson: efficiency required attention to costs as well as benefits:

> The basic issue is not whether DHEW has the authority to establish quality control systems, or whether the imposition of penalities has a supportable statutory basis. The real issue is that the 'proper and efficient operation of the state plan' in the fifty-four AFDC jurisdictions cannot divorce the management function of detection, correction and the prevention of erroneous payments from the associated problem of adequately controlling the cost of program adiministration. The Quality Control Program has established an artificial and malignant dichotomy by setting arbitrary controls subject to punitive sanctions without regard to what is essentially a trade-off decision making process between error costs and administrative expenditures . . . . Reductions in ineligible and overpayments will become progressively more difficult to achieve as the 3% and 5% tolerance levels are approached or to maintain, once achieved, so that increasing applications of resources will yield decreasing returns, thus escalating administrative expenditures to proportionally higher plateaus in relation to error costs. (Committee on Ways and Means, 1976, pp. 4, 9)

In fact, efficiency, which program officials understand and care about much less than their frequent use of the term implies, may not even have been what they had in mind. As one analyst points out (Mendeloff, 1982), their real intention was about something they both know better and care a great deal more about, cost control:

> Conspicuously absent from HEW's new guidelines is any emphasis on *underpayment*, the third error rate that auditors traditionally report; in California it was 6.7%. The omission is intentional . . . . Overpayments impose costs on the federal government, the states and the counties; underpayments, on the other hand, are costly only to the affected recipients and represent savings to government at all levels. (p. 262) (emphasis in original)

The outbursts of administrative energy that often characterize the promulgation of new regulations can serve still other purposes. For example, they may help neutralize political pressure to "do something" about a problem, as the beleaguered head of a troubled state program for the elderly found:

> Even though they didn't like what we were doing a lot—would have done it differently, more of this and less of that—there wasn't really any debate about how things were going to get done. Everyone pretty much bought into it. Emma (omnipresent, long time client advocate) may have though we were full of crap, but she wanted new regulations, by god. If (Senator) Garamendi thought we were full of crap, he wanted program (rule) revision. DO SOMETHING ABOUT IT! We got daily confirmation of that. (Pruger, 1987, p. 207)

In sum, the relation between top down regulatory efforts and efficiency, in whose name they are almost always taken, is uncertain at best. Nevertheless, over and over again the connection between the concept and the imposition of irrational standards is reenforced. Sometimes the complaint is a relatively mild one: "Most of us have experienced the frustration . . . of struggling to interpret and implement new procedures we had no part in developing, no chance of questioning, and of whose value we had little conviction" (Watson,

1983, p. 20). However, that "little conviction" readily escalates to evasive game playing and deep resentment from below, making the problem even more intractable (Weatherly, 1960). This pushes even further into the future the time when efficiency might guide the production and distribution of social services.

4. Efficiency is not a lack of concern for quality. The breakup of the Bell System caused problems with the local telephone company's directory assistance service. As reported in the press, a company spokeswoman acknowledged this with the remark, "When you are under the gun to be competitive, the first thing you do is increase your efficiency. If you strive for efficiency, sometimes quality and courtesy suffer" (*San Francisco Chronicle*, August 29, 1988). In response, a letter from a reader bitterly complained, "(The spokeswoman's statement) is both a frightening comment on our times and a foreboding omen for the future . . . . The mind boggles at the idea that a service industry, which has only itself to sell, can exclude 'quality' and 'courtesy' from its understanding of 'efficiency'" (*San Francisco Chronicle*, September 19, 1988).

Nevertheless, it is precisely this exclusion that is frequently made in the social welfare literature. Buxbaum (1985), among many others, shares the misconception:

> The prime objective (of cost-benefit analysis) is efficiency. As pointed out this is an important objective; without it the pie shared by each member of society diminishes and everyone receives a smaller slice. The size of the pie, however, is not our only concern. Quality is also a factor. (p. 71)

Anyone schooled in the concept knows that efficiency includes *all* the dimensions of the output or outcome, including the qualitative ones. When any dimensions are excluded the fault is in the application and/or the evaluator, *not in the concept*. Thus, the headline that introduces the reader's letter—"Quality, Efficiency Go Hand in Hand"—makes the correct point. Conversely, it is an incorrect use of the term to say, "An increase in efficiency, if it is achieved by lowering quality. . . ." (Gates, 1980, p. 176). (In other places, however, Gates uses the correct meaning. For example, ". . . in evaluating efficiency, decision makers . . . must ensure that each unit of service being compared is of equivalent quality'' (p. 182). The fact that, here again, a sophisticated writer is using different meanings of efficiency underscores the necessity to bring as much clarity to the matter as possible.)

A profit making telephone company that claims greater efficiency because it ''trimmed three seconds to four seconds off the average call time'' may, in fact, have created greater inefficiency. A nonprofit or public social service agency that makes the same claim based on an increase in the number of cases processed by each worker is just as likely to have made the same error.

''Quality,'' when used in a discussion of social service objectives and effects, serves several purposes. One of those is its use as a catchall term for outcomes that are hard to measure (e.g., diminutions in the warmth of the service giving transaction that often follows an increase in worker caseload). Because they are hard to measure, they are not measured. This makes them especially vulnerable to being innocently overlooked or consciously, even cynically, sacrificed during times of programmatic adjustment to shrinking resources.

Another use refers to effects that the decision making authority considers secondary. Thus, it is willing, albeit reluctantly, to sacrifice them when conditions are felt to require it. This authority will be whoever is responsible for the budget. Its decision will most likely not reflect the views or values of the staff, whose chief responsibility is the delivery of services. For example, California's program of in home supportive services for aged and disabled persons originally promised whatever services were needed to keep clients in their own homes ''in safety and comfort.'' When cost control became the pressing concern this pledge was reduced to provide only for ''safety.''

Patti's (1985) list of examples is made of both kinds: ''Accessibility, timeliness, consistency, humaneness, and technical proficiency of services,'' he says, ''are examples of quality'' (p. 8). What is relevant here is that losses of both kinds are regularly, but carelessly, associated with the pursuit of efficiency. The cause of the losses, and therefore the solution, lie elsewhere. Regarding the first kind, some way needs to be found to better represent hard to measure but desirable program effects. There seems little that is progressive in insisting that there is too much nuance here for this:

> One hears that programs of social work produce outcomes so subtle they cannot be measured but that somehow without them society would be worse off. When the outcome is not measurable, social workers are probably engaging in self-delusion. (Newman & Turem, 1985, p. 169)

Losses that occur because the decision maker considers certain outcomes expendable are harder to address. Shrinking resources sooner or later impose shrinking aspirations and accomplishments. Service staff involvement in the tradeoff deliberations might help. And genuine increases in efficiency might make those tradeoffs less bitter. In that sense efficiency criteria could lead to the *preservation* of quality. In any case, it is incorrect to connote a necessary relationship between efficiency and the diminution of service quality. The following excerpt makes that error. (To emphasize the point being made, we have inserted the needed correction; it is italicized and contained within parentheses):

> Wherever there is some slack in the definition of social welfare program objectives and uncertainty around performance measures, (*improper or opportunistic use of*) the standard of efficiency perforce pushes in the direction of minimal provisions: child development programs slide toward routine babysitting operations, planned parenthood toward pill dispensing, nursing homes toward human warehousing, and so forth. (Gilbert, 1983, pp. 28-29)

5. *Efficiency is not antagonistic to innovation*. Activities that make up the search for innovation are paid for in two ways: outside the normal budget and within it. In the first instance, funds are made available *only* if the experiment is undertaken. They cannot be used to advance other organizational ends, including improved service giving. Thus, they neither increase nor decrease the resource pool available to achieve current agency objectives. As such

they are irrelevant to a judgment of the efficiency of the ongoing operation.

Innovation can also be funded from the agency's operating budget. In this circumstance, more experimentation necessarily means less service giving, and *vice versa*. Here, because resources are scarce, efficiency criteria *are* relevant. As one way to state it, efficiency dictates that the experiment be funded if the expected gain is deemed more valuable than the services that would now not be offered. If the reverse is true (i.e., the foregone services are considered more important than the expected contribution of the experiment), it would be efficient *not* to fund the experiment.

The central point is that efficiency does not prejudice the case regarding innovation. Nevertheless, demonstrating one more misuse of the concept, the social services literature frequently portrays efficiency as the enemy of innovation:

> The greater the degree of efficiency, the lower the rate of change. . . . Because innovation is costly, it may not be encouraged in an organization concerned primarily with efficiency. (Neugeboren, 1985, p. 202)
>
> In their quest for greater efficiency, social agencies have assumed that time devoted to rethinking ideas already clearly formulated – or spent reexamining practices that have already proven themselves effective – is time wasted. (Watson, 1983, p. 19)

One other similar circumstance deserves mention here. Some problems are seen as so important that the decision maker considers even small gains as being very valuable. This justifies what otherwise seems to be disproportionately large resource allocations. An efficiency calculation can encompass such valuations, but it cannot supply them.

The relative valuations must come from whomever is authorized to determine them. If more than one valuation and decision maker are involved, only politics can resolve the conflict. For example, some do not believe in spending critically scarce resources on problems already shown to be very hard to solve. This view may win out over the opposite view. The loss here is properly attributable to politics, not efficiency. As Reisch and Taylor (1985) recognize, however, common usage implies that the reverse is true:

> Programs which address the most intractable problems in our communities may be the least 'efficient' in cost-benefit analysis terms and become the first to be cut in period of retrenchment. (p. 299)

CONCLUSION

Efficiency is not cost cutting; nor is it being businesslike. It does not mean a lack of concern for quality; nor does it lead to rule by irrational standards. It is not a drag on creativity, innovation and experimentation; nor is it advanced as a substitute for politics. Yet, this is precisely what many in social welfare – academics and practitioners alike – believe, or at least carelessly imply.

Properly understood and used, efficiency would work against the unreason that is an aspect of all of the above conditions. This potential is virtually lost to the social services enterprise, where the concept is treated either as worthy of contempt or a tool of minimal importance applicable only to relatively simple, uninteresting circumstances.

To be sure, there are substantial intellectual barriers that block the widespread, informed application of efficiency to the production and delivery of social services. This paper has only discussed the trivial ones. Many believe them, and that is why they had to be addressed; there is, however, no conceptual substance in them. Some of the weightier matters are considered in Part B of this paper.

REFERENCES

Alger I. (1980). Accountability: Human and political dimensions. *American Journal of Orthopsychiatry*, *50*, 388-393.

Buxbaum, C. (1985). Cost benefit analysis. In S. Slavin (Ed.), *Managing finances, personnel and information in human services*. New York: The Haworth Press.

Committee on Ways and Means, Subcommittee on Oversight. (1976). *Hearings: HEW efforts to reduce errors in welfare programs (AFDC and SSI)*. Washington, DC: U.S. Government Printing Office, April 29 and May 3, 1976.
Dery, D. (1981). *Computers in welfare: The mis-match*. Beverly Hills: Sage Publications.
Fabricant, M. (1985). The industrialization of social work practice. *Social Work*, *30*, 389-395.
Gates, B. (1980). *Social program administration: The implementation of social policy*. Englewood Cliffs, NJ: Prentice-Hall, Inc.
Gilbert, N. (1983). *Capitalism and the welfare state: Dilemmas of social Benevolence*. New Haven: Yale University Press.
Good, P. (1987). Brief therapy in the age of reagapeutics. *American Journal of Orthopsychiatry*, *57*, 6-11.
Hasenfeld, Y. (1984). The changing context of human services administration. *Social Work*, *29*, 522-529.
Henderson, R., & Stone, B. (1974). Accountability for what and to whom? *Social Work*, *19*, 387-388, 507.
Judge, K., & Knapp, M. (1985). Efficiency in the production of welfare: The public and the private sectors compared. In R. Klein & M. O'Higgins (Eds.), *The future of welfare*. Oxford: Blackwell.
Knighton, A., & Heidelman, A. (1984). Managing human service organizations. *Social Work*, *29*, 531-535.
Lohman, R. (1984). Resource development as executive leadership. In F.D. Perlmutter (Ed.), *Human services at risk*. Lexington, MA: Lexington Books.
Mendeloff, J. (1984). Welfare procedures and error rates: An alternative perspective. In J. Tropman, M. Dluhy, & R. Lind (Eds.), *New strategic perspectives on social policy*. New York: Permagon.
Mullen, E. (1984). Book review *(Verdicts on social work)*. *Social Work*, *29*, 309.
Neugeboren, B. (1985). *Organization, policy, and practice in the human services*. New York: Longman.
Newman, E., & Turem, J. (1985). The crisis of accountability. In S. Slavin (Ed.), *Managing finances, personnel and information in human services*. New York: The Haworth Press.
Niskanen, W. (1971). *Bureaucracy and representative government*. Chicago: Aldine.
Patti, R. (1987). Managing for service effectiveness in social welfare: Toward a performance model. *Administration in Social Work*, *11*(3/4), 7-22.
Pruger, R. (1987). *Making information work: The story of the equity project*. Unpublished manuscript.
Reisch, M., & Taylor, C. (1985). Ethical guidelines for cutback management. In S. Slavin (Ed.), *Managing finances, personnel and information in human services*. New York: The Haworth Press.
Simon, H. (1987). The proverbs of administration. In J. Shafritz & J. S. Ott (Eds.), *Classics of organization theory*. Chicago: Dorsey Press.

Taubman, S. (1985). Doing less with less. *Social Work*, *30*, 180-182.
Watson, K. (1983). Efficiency versus process. *Public Welfare*, *41*(3), 19-23.
Weatherly, R. et al. (1980). Accountability of social service workers at the front line. *Social Service Review*, *54*, 556-571.
Webster, N. (1961). *Webster's new international dictionary of the English language*. 2nd ed., unabridged. Springfield, MA: C.C. Merriam, 819.
Wildavsky, A. (1976). The strategic retreat on objectives. *Policy Analysis*, *2*, 499-526.

Efficiency and the Social Services: Part B

Robert Pruger, DSW
Leonard Miller, PhD

Economists "own" the efficiency concept. But they have never made a persuasive case about its applicability to the social services. (The best try we are aware of is Williams and Anderson, 1975.) Indeed, one economist with long exposure to social workers, concluded that her discipline could hardly be of less value to the direct service practitioner: "How much help can the social worker get from economics? I am afraid that the answer is 'not a great deal.' . . . For the clinician, it seems to me that economics has almost nothing to offer" (Burns, 1974, p. 103).

Burns and her colleagues come to the opposite conclusion regarding non-clinical personnel, but even here the case is not a strong one, and ultimately it rings hollow because it misses the core of what social workers actually do. For example, it is one thing to argue that at least policy specialists should become familiar with a few, well-selected economic ideas:

> For the social policy practitioner economics has much greater relevance . . . the concept of cost in the sense of opportunity cost can make the social policy practitioner a much more *efficient* and responsible planner and deviser of programs, less likely to be brushed off as an unrealistic or ivory-tower dreamer. (Burns, p. 104) (emphasis in original)

But it only invites incredulous, glassy-eyed stares to insist:

Dr. Pruger and Dr. Miller are Professors, School of Social Welfare, University of California at Berkeley, Berkeley, CA 94720.

> Social workers must do more than read the economic critics of the system—for instance, the Galbraiths and the Bouldings. . . . Public policy social workers must understand the difficult and technical work of the Arrows and the Lampmans. (Page, p. 52)

> (The social worker generalist) needs to know the tenets of fiscal policy including an introduction to both taxation and expenditure policy. His study should also cover the Keynesian theory of the relationship between national income, employment and prices. (Tucker, p. 100)

The last two citations show a greater knowledge of economic theory than they do an understanding of social work's role in the production and delivery of social services. In any case, to make the case for efficiency in terms of fringe groups (e.g., "public policy social workers," "social work generalists") is to prove its irrelevance to the profession as a whole.

A proper argument requires a connection between the concept and common duties. The efficiency perspective instantly draws attention to one such feature. To wit: social workers at all hierarchical levels make discretionary decisions about how scarce resources are used. Efficiency can frame and inform these choices.

Though this connection is straightforward and strong, it remains to be shown that the concept can be usefully applied. It is after all, reasonable to wonder: How can microeconomics—a theory of efficiency based on markets, prices, the profit motive and competition—be applied to the social services where, in the normal instance, all of these conditions are absent? This question is taken up immediately after making a point about the audience our argument is intended to address.

Some respond to microeconomic theory not as a set of ideas to be considered but as an ideology to be opposed. They see it as threatening cherished values. Thus, when one economist recommended the overall perspective to social workers, Aigner and Simons (1977) sounded the tocsin:

> What Page has offered to social workers as a useful model has little relevance for our profession. The microeconomic model of pure competition is based on assumptions antithetical to the values that underpin social work . . . . Rather than requiring a commitment to abstract economic principles, the future demands a commitment to humanism, to a view of people as something more noble than mere seekers of profit and to the wise use of resources for community survival . . . . We must faithfully implement our primary and instrumental values in our practice. Otherwise, we may unwittingly lose our souls. (pp. 307-308)

We will not try to bridge the gulf between those for whom microeconomics is ideology and those for whom it is an analytic perspective. Our argument is directed at those willing to consider the ideas but skeptical about their applicability.

EFFICIENCY WITHOUT COMPETITION

Even people who have never studied microeconomics have very serviceable notions about what efficiency is: for example: "getting a bigger bang for the buck"; reaching the objective at the lowest cost. Most also know that efficiency is an end, while things like competition, markets, and self-interest are means to reach it.

This important distinction gets blurred because the means are often treated as the ends. The solution to many policy problems, for example, is stated in phrases such as "make the system more competitive," "strengthen the market," and the like. The only rational basis for achieving these conditions, however, is that they are presumed to lead to greater efficiency. There is no reason why human beings should want more markets or competition, *per se*; there is every reason to expect them to want more value out of their limited resources.

All introductory microeconomics texts offer the theoretical analysis that explains why things like markets and competition are so powerfully linked to efficiency. Indeed, that connection is so strong that it leads almost all nonspecialists to misunderstand the "message" that is in microeconomics. That message comes in two parts:

first (as everyone knows), competition and its associated conditions are means to achieve efficiency; but second (contrary to what most believe), *they are not the only route to that end*. Efficiency, in brief, does not depend on them.

Examine a typical expression of the economists' fundamental but odd-sounding understanding of efficiency. It immediately becomes apparent that more than "getting a bigger bang for the buck" is involved (though that meaning, too, is encompassed):

> A society has allocated its resources efficiently when no input and no output can be transferred to some alternative use without making at least one person worse off, even if the welfare of others is improved. (Culyer, p. 22)

Note that the focus is entirely on a set of resource-allocating decisions. There is no mention of a preferred or necessary process or institution for making those decisions. The decisions are measured only against a standard of perfect efficiency, which is reached when there is no further scope for mutually beneficial exchange. This is the most efficient condition possible because, given the preexisting distribution of resources, everyone has been made as well off as he or she possibly could be—without making someone else worse off. The familiar story is that the dynamics of a perfectly competitive market lead to this result. The less familiar story is that there are other ways to come to the same result.

An alternative mechanism resembles how decisions are made in bureaucracies, a form much closer to what one finds in the social service enterprise. One person or group in authority determines what should be done, and this message is communicated. Some call this allocation by command; others call it planning. If the resulting allocations are the same as those that would result from using perfectly functioning market institutions, then the plan, too, is perfect. Accordingly, *a perfectly planned economy (one without competition) and a perfectly competitive one (one without planning) would be equally efficient*. Both would get the most value from the resources at their disposal.

Self-interest motivates actors in a competitive environment to discover and make the most efficient allocations. But self-interest has no place in a theoretically perfect planned economy. Such economies do, however, readily accommodate an alternative allocating mechanism – professional ethos. The beliefs, habits and values that characterize professionalism qualify it as an institutional alternative to competitive self-interest.

Professionalism involves a commitment to rational behavior. Thus, when the decision makers are professional and when they know what decision would lead to the most efficient use of resources, they need neither markets nor hierarchical direction to achieve that result.

All of the above, of course, has been stated in ideal terms. Perfect competitive markets do not exist, nor do perfectly planned bureaucracies. Nor is professional ethos the only thing that guides the behavior of professionals. Nevertheless, barriers to perfectly functioning markets are better understood and prescribed for than are barriers to perfectly functioning, professionally staffed bureaucracies. Much more than can be said here is needed to show how the economist's efficiency criterion can be comprehensively applied to the agencies and programs that we call the welfare state. The statement that microeconomic theory does not tie that accomplishment to the creation of competitive markets is simply the first important clarification.

Next, we consider one of the barriers: the prevailing understanding of efficiency and effectiveness. The misconception found here helps explain why social workers do *not* seek to learn what decisions would lead to the most efficient use of resources.

THE CONTINUUM OF RATIONALITY

Service effectiveness is regularly advanced as the highest criterion of accountability in the field: "The ethic of service validates intense concern for the effectiveness of social agency services, and ultimately this is our overriding concern" (Gruber, 1986, p. 13). Patti (1987) goes even further; he specifically ranks all other standards as secondary:

> The main (and perhaps too obvious to mention) reason why service effectiveness should be the principal concern of management is simply that this is the primary business of the social welfare organization. Changing people and/or the social conditions in which they live is the *raison être* of the human service agency, not the acquisition of resources, the efficient use of resources or the satisfaction and development of staff. All these may be important in their own right, and even instrumental to providing effective services, but they are, or at least should be, subservient to this objective. (p. 9)

Neugeboren (1985) carries the point one step further still. Effectiveness, he warns, can be sacrificed to efficiency, and he promises a way to avoid this:

> The pressure for better utilization of existing resources has thus raised the issue of the relative importance of, and the relationship between *efficiency* and *effectiveness* of social programs . . . . Might not the emphasis on efficiency and productivity have a negative impact on the effectiveness of social welfare, that is, on the extent to which agencies can serve client needs? An approach is suggested here that may help avoid the substitution of efficiency by emphasizing social policy goals as an integral part of human service practice. (pp. 3-4) (emphasis in original)

Both citations treat effectiveness and efficiency as being entirely different things, illustrating a difference *in kind*. Thus, not only can they be pursued separately, but one can even choose to advance one and ignore the other. Complementarity is possible, but there is a stronger presumption of conflict; and when they conflict professional values dictate that effectiveness be served before efficiency.

This view of the character of the difference between the two concepts is very widely held: "Nonprofit organizations give (efficiency) a lower priority than effectiveness" (Elkin, 1985, p. 147); "Efficiency measures . . . displace emphasis on effectiveness" (Hasenfeld, 1984, p. 527); "Efficiency and effectiveness may be traded" (Gruber, 1986, p. 7). Though common, the view is seriously flawed. Efficiency and effectiveness are different, but something more like the difference between hot and cold is involved, i.e., a difference *in degree*. An entirely different perspective and conclusion follow from this.

Every point in the search for better ways to reach social service objectives has *two* dimensions: some amount of effectiveness *and some amount of resource expenditure*. At the zero expenditure level, for example, the effectiveness level also is zero. Correspondingly, greater levels of accomplishment can more easily, and sometimes only, be accomplished at higher cost levels. A constant challenge is that it is always possible to achieve a better effectiveness per dollar ratio than the one that exists.

Interest in achieving an improved ratio is demonstrated in two ways. First, one can try to be more effective at the same resource cost (in the economists' jargon, "maximize output subject to a budget constraint"). Or, one can try to reduce costs without reducing effectiveness ("minimize cost subject to an output constraint"). Note that both costs and effectiveness are always encompassed in the way the problem is understood, no matter what words are used to express it. From this perspective, it is impossible to trade off effectiveness against efficiency and, therefore, pointless to rank one as more important than the other.

Consider now a continuum in the number and types of considerations about how to achieve the rational administration of social service resources. As one moves from fewer to more considerations about this question, one moves from a more "bounded" rationality to a less "bounded" one (Simon, 1957). Because fewer considerations are taken into account about effectiveness it is more bounded in its rationality. Efficiency, involving a wider range, is less bounded. Such a continuum can be drawn as shown in Figure 1.

Figure 1

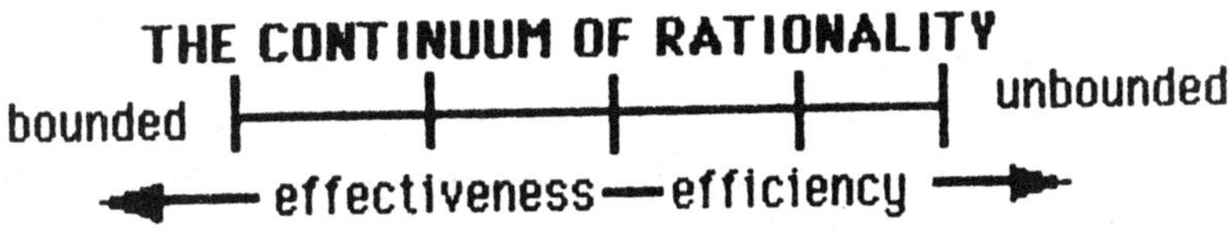

Any point or interval on this continuum is best understood by the number and types of questions asked to inform rational program improvement. The presentation below is ordered by the central question that exists at selected intervals. As one moves along the continuum, the supply of information grows, and with it, the possible scope and precision of rational action.

Does the Program Have a Desired Effect?

At one end of the continuum, where rationality is in its simplest form, all elements of the situation are treated as unchangeable except the grossest one – the helping activity as a whole. The only evaluative question posed is: Do or do not clients experience a program's effect? With the answer in hand, the search for improvement can proceed, but it can be no more precise than that single piece of information permits: programs that have effect can be implemented, and those that do not can be rejected.

Given That the Program Is Effective, What Is the Size of the Effect?

The next significant interval along the continuum incorporates a quantitative measure of outcome change. Here the evaluative question is, how much of each program effect is associated with each intervention? The answer makes it possible to choose between more and less effective efforts. As Reid (1987) states it:

> The establishment of effectiveness, while fundamental, is only the beginning. Managers and their constituencies want to know more than whether or not a program is effective. They want to know 'how effective' it is. (p. 44)

Many empirical investigations of practice yield information of this sort. For example, Reid and Hanrahan (1982), Thomlinson (1984), and Rubin (1985) review the change in clients attributable to various clinical or therapeutic interventions (e.g., the comparative effects of conjoint treatment and communication training in marital counselling). Kemper, Applebaum, and Harrigan (1987) describe how much various case-managed community care programs for the impaired elderly impacted outcomes of concern (e.g., nursing home and hospital use, informal care giving, client quality of life).

Given That the Most Effective Program Is Being Used, with Whom Is It Most Effective?

Every program has to answer the question: Who should be the client? The answer has two parts. The first is provided by political authority: it determines who is eligible. Thereafter, as Newman and Turem (1985) point out, rational information becomes relevant:

> Basically, eligibility requirements tend merely to reduce the universe of potential users rather than select among potential clients those whose utilization does most to accomplish the program's ends . . . . The first-come, first-served technique tends to be inefficient in that it does not identify the range of clients from whom those who would use the service resource most efficiently might be selected. (p. 163)

Sometimes this is referred to as the targeting question. Johnson's (1984) research, for example, asks: If the child protective services unit does not have the resources to adequately serve all child abuse victims, which families are most likely to show further episodes, and therefore should be attended to first?

In another service arena, community care programs for the elderly, the targeting or "for whom" question is generally considered the key to their effectiveness. For example, Miller (1988) estimates that the average effect of one such program on the most frail group was almost 20 times what it was in the stronger group. Weisseart, Cready, and Pawelak (1988) reviewed 31 community care programs, approximately half of which evaluated nursing home use for subgroups of clients. The detailed results for various subgroupings (i.e., physically disabled clients, socially deprived clients, clients covered by Medicaid and on a nursing home waiting list) provided little direction. However, when one of the largest programs estimated the risk of institutionalization, a treatment effect was found where one had not existed before. The researchers conclude, "Future studies might be best served by defining subgroups on the basis of estimated risk of institutionalization" (p. 329).

Given That the Most Effective Program Is Being Applied to the Most Productive Client Situations, How Much Service Should Any Given Client Receive?

To say that one program is more effective than another does not mean that it is more effective in every instance, only *on average*. Information about average effects is of no help to the social worker who must decide client treatment on a case by case basis. The essential question one faces is: What is the right amount of service to award to the client before me? This is the matter addressed at this point on the continuum.

The search for the needed answer requires a conceptual and methodological shift. This is from concern about average effects of resources (or average effects per subgroup), to concern about marginal effects of their use—the extra effect on a client that follows from an extra amount of any particular activity or service.

The central idea here might be stated as follows:

1. Every unit of service, say an hour of the worker's time, has the same cost.

2. The effect on the client for each worker time increment, however, is not the same. Up to some number of hours, the effect per hour may grow. Beyond that point the effect of each succeeding service hour will decline. There always is a "point of diminishing marginal returns."

3. Marginal resources should be applied to the client with the maximum effectiveness per hour (which amounts to the maximum effectiveness per dollar). The continued application of this rule will cause the effectiveness of the last hour (dollar) of resources applied to all clients to be the same. When resources are applied to clients past this point, rather than those not yet at it, the program is being less effective than it could be.

4. Information collected at previous points on the continuum made it possible to offer to the most appropriate clients the services expected to be the most effective. The next increment in rationality requires that each of these clients be given the most productive or optimum amount of services. The information called for in this re-

gion of the continuum is what the program needs to accomplish that result.

The concept of marginalism is a major tool for economists: "The elaboration of (marginalism) transformed economics in the late nineteenth century, and the fruits of the marginalist revolution continue to set the basic framework for contemporary microeconomics" (Rhoads, 1985, p. 25). But most human service personnel have little or no working knowledge of it.

Social workers are familiar with statements about aggregate effects. These take the form, "X treatment produced Y result." The absolute amount of Y is usually the focus of attention. They are much less familiar with statements of marginal effect. These take the form: "the first unit of X produced this fraction of Y; the second unit of X produced this next fraction of Y; and so forth for all increments of X." Here the most important thing to observe is the amount of each successive increment of Y.

If line social workers know the optimum amount of service to award to each client, the program as a whole can realize more of its potential. The in-home care program with a deinstitutionalization objective, the alcoholism clinic with a detoxification objective, or the foster care program with a family reunification objective, can achieve the most their service budgets allow.

Unfortunately, the information workers need to make these decisions in the most rational way is virtually never available. Few recognize the need for it, and even fewer in the field have the training required to produce it. Consequently, this point along the rationality continuum identifies a future rather than a current possibility.

Extending the Continuum and Filling in the Gaps

All the information collected thus far optimizes a static situation. But over time everything will change: some client needs will grow and others diminish; worker productivity will improve or worsen; budgets and costs will rise or fall; standards, codes, laws and other externally imposed conditions will impose new constraints and open new opportunities; and the like. Modern information technology can powerfully augment the program's capacity to monitor and respond to these dynamic factors, as well as learn from its own practices over time. The most unbounded region of the rationality continuum, therefore, is concerned with how to optimize under ever changing input, output and environmental conditions.

The continuum can also be specified better along its full length by clarifying intervals passed over in the above sequence: What caseload size is optimum? Does the program have the right incentive structure? Are costs correctly understood and measured? If two or more services contribute to the same end, are they being used in the most cost-effective combination? What is the efficient allocation of effort if two or more objectives are being pursued through the same service (e.g., in-home services to increase client quality of life, as well as reduce the probability of need for a nursing home)?

In any case, the idea of a continuum of rationality is more fundamental than any of its details. To tie that down, it may help to imagine that all the information needed to improve a program was wrapped in layers around a core. Some layers lie closer to the surface and some closer to the core. The questions asked at each successive interval in the continuum address an exposed layer. The answers obtained peel that layer away, creating another piece of the intelligence needed to further improve the program, and exposing a previously hidden layer to examination. In this way, each increment in evaluative information transforms additional elements of the helping technology into variables, elements that at previous points along the continuum are treated as fixed. Social service practice and research have penetrated the topmost layers; the continuum of rationality challenges the enterprise to reach deeper into the puzzle.

In this section, we have argued that the prevailing view of effectiveness and efficiency is inadequate. Conceiving the primary professional obligation to be effectiveness misses the inescapable reality that every step taken to discharge it necessitates the expenditure of scarce resources. For that reason alone (and there are others) the field needs to expand its understanding of its fundamental obligation to include the best possible use of those resources. The continuum of rationality, along which effectiveness and efficiency are merely the names given to the poles and movement toward the efficiency end is the proper commitment, is the appropriate conception with which to begin.

THE PROBLEM OF GOALS AND OBJECTIVES

Over the years scholars have regularly portrayed organizational and program objectives as incapable of guiding rational action. They found them to be "invariably . . . multiple, conflicting and vague" (Neinberger & Wildavsky, 1973, p. 10); "multiple . . . and conflicting" (Gates, 1980, p. 225); "impalpable and multiple" (Gilbert, 1983, p. 11); "inconsistent, contradictory, or incoherent" (Kanter & Summers, 1987, p. 155). One might wish that these problematic characteristics were mere errors, for errors can be corrected. But their source is much deeper—and much less changeable. They reflect the condition of the different values and perspectives that swirl around a program at all levels.

While rationality cannot resolve such matters even badly, politics does so routinely. Among other ways, it accomplishes this by converting defects into virtues:

> Fuzziness is also inherent in many goals because their very vagueness and ambiguity serve important political functions. Under the umbrella of pietistic generalities, diverse interests can form working alliances as if there were agreement . . . . The imperative to state goals in clear and precise terms . . . is an arguable matter. (Gruber, 1986, p. 3)

Political requisites can even turn things upside down, making goals not the first but the last thing considered. "Some goals may be . . . defined *post hoc* by the organization to justify its actions" (Kanter & Summers, 1987, p. 155). A more common distortion is that, "The political process often gives more weight to impacts that are concentrated, tangible, certain, and immediate than to impacts that are diffuse, intangible, uncertain and delayed" (Weimer & Vining, 1989, pp. 199-200). Even what "is often viewed as a measurement issue (has) its roots in conceptual and value orientations" (Reid, 1987, p. 46). In these and other ways goals and objectives are shaped by politics, values and ideology more than the requisites of rationality. And it will forever be so. Thus, one must wonder. . . .

Wither Efficiency?

What matters most is not the reported defects, *per se*, but the meaning given to them. That is, are they to be taken as evidence of the impossibility of rational administration, of applying efficiency to the social services? Or simply as suggesting the institutional complexities and limits within which rational progress will have to be made? The latter, which we urge, has a far stronger defense.

To begin with, those who declare efficiency to be an unusable standard of evaluation, must also reject effectiveness, since both depend on the same information about objectives. And to reject both is to take the enterprise beyond accountability. On the other hand, endorsing the efficiency standard does not require arguing that it can be productively applied in all circumstances. There is, for example, no need to insist that every possible goal can be quantified without worrisome distortions. It is sufficient that rational criteria—effectiveness and efficiency—are applicable across a wide range of social service situations, and with further technological and conceptual advances that scope can be increased over time. We conclude this paper with a brief review of some of the reasons for this belief:

1. The problem of multiple goals is manageable. Rationality does not require that all objectives and outcomes be considered all the time. Agreement among the parties involved or political reality often mitigate the problem of multiple goals: they make a very few objectives, or even a single one, the major focus of attention. Regarding workfare, for example, it is entirely valid to ask things like, do parent-child relationships improve or worsen when welfare mothers are compelled to accept a work obligation? But everyone knows that the commanding objective here concerns the program's impact on relief rolls and costs. Davies and Knapp (1988), in reporting on the work of the institute they lead, further exemplify how a program's recognized constituencies help define the primary outcomes of concern: "Undoubtedly, the questions on which we have done most work are what are the costs of the outcomes which would be considered by interest groups important to the evaluation of spending on care" (p. 2).

Reduction of objectives to enhance the possibility of rational management is also arranged through administrative leadership and discretion:

> Instead of setting dozens of objectives, as the Nixon administration did, Bush is urging his managers to choose a handful. In his memo, the President wrote: 'I request each of you to identify three to five candidate objectives that you believe should be major priorities in your area.' . . . The memo went on to say that OMB, in conjunction with the White House Office of Policy Development, would help Bush select 'two or three objectives in each of your areas.' (*San Francisco Chronicle*, April 17, 1989)

2. *Some important objectives and outcomes are unambiguous and easy to measure*. Rapp and Poertner (1987) identify one type:

> Client status change is an inherent component of many service systems. Examples include: a child is removed from his/her home, a case is opened, a case is closed, a person acquires full-time employment, or a person moves from an institution to a group home. Recidivism is a common concern in areas such as mental health (return to the hospital) and corrections (return to prison). A cursory glance at these examples suggests that they have few of the measurement problems which plague other client outcomes. (p. 31)

Reid (1987) identifies another type:

> In some programs, changes following interventions bear an obvious imprint of the intervention. For example, a program is designed to teach retarded children specific interactive skills—e.g., learning to initiate a conversation. Observation reveals that these skills, absent before the program, are present following it. (p. 43)

3. *Technological improvements already make it possible to define and measure a wide array of subtler or more complex outcomes*. If that were not so, Giannetti and Wells (1985), for example, could not find "considerable . . . data on the differential effectiveness of the three major mental health professions of psychi-

atry, psychology and social work;" nor could Hanrahan and Reid (1984) report "an impressive empirical basis for the claim that a variety of methods in the helping professions can achieve beneficial results across a broad range of client populations"; or Reid (1987) to cite "recent work on quality assurance approaches (that) suggest specific procedures for assessing this type of effectiveness that go far beyond familiar but crude process evaluations (e.g., number of client contacts)." Indeed, Judge and Knapp (1985) are almost categorical in their insistence on the point:

> It used to be suggested that output measurement was too imprecise, but this is no longer a convincing argument. Conceptual and empirical developments in the multi-dimensional measurement of well-being and health status, for example, facilitate the careful design of research instruments to collect data with which to model sophisticated simulations of the production of welfare. (p. 146)

4. A proper analysis of social service objectives and of the relationship of the usage of resources to achieve them could lead to substantial efficiency gains. Many social service objectives are, as reported, "impalpable." However, this does not mean that they are incomprehensible, merely "not perceptible to the touch; too slight or subtle to be grasped easily by the mind."

Impalpable objectives *can* be rationally pursued, but it would take a particular vision and appropriate knowledge to do so, things social service managers usually lack. Thus they, especially when under fiscal or political pressure to show results, settle for what they can see, for what is "perceptible and *not* too slight or too subtle to be grasped easily by the mind"—things like the count of clients seen and cases closed. Academics criticize this practice, but so long as administrators have no understanding of the science of scarcity of social services they will continue to rely on management techniques adequate to their limited perspective.

When Reisch and Taylor (1985) ask, "How can administrators apply a utilitarian calculus to degrees of suffering and deprivation?" they are asking about the measurement of one of the most impalpable and fundamental of all social service ends, care. Such questions are often raised in a way that connotes that too much subjectivity is involved for *any* answer to be authoritative. Nevertheless, the problem does have a theoretical, rational solution (Miller & Pruger, 1988). In addition, a technology for realizing the solution has been successfully implemented (Pruger, 1987).

The data analysis that informed this work demonstrated that what can be measured—i.e., the relationship between worker awards and client need in a care program—contained all that had to be known about what cannot be measured—i.e., the social worker's subjective evaluation of each client's well-being. In brief, there *is* a utilitarian calculus that can be applied to degrees of suffering and deprivation. Moreover, it can be done in a way that both honors the necessarily subjective content of the worker's decisions, but still results in an equitable, efficient allocation of the award budget.

There is a second fundamental social service end; we have elsewhere called it "people changing" (Miller & Pruger, 1989). It is concerned with helping people accomplish any of an infinite variety of objectives: become employed; overcome marital problems; live successfully in the community after a stay in a mental hospital; and the like. The key point here is that professional effort cannot make these things occur; all it can do is increase the probability of their occurring. This idea is not broadly understood in the field, nor are the operational and methodological requirements that follow from it. Doctoral programs in social welfare, for example, do not appear to offer the requisite instruction. On the other hand the underlying theory and evaluative techniques do exist. With the appropriate research effort, they could be adapted to help rationalize the production of people changing services.

In sum, there are two elemental social service outcomes. The first, care, is impalpable and subjective; the second, people changing, is impalpable and objective. Both can be rationally accounted for. An appropriate analysis of these ends, along with the corresponding development of the methods needed to evaluate them, would go a long way toward making the concept of efficiency familiar and even welcome in the field.

A PERSONAL CONCLUSION

As the available opportunities and our alternatively rising and sagging morale permitted, we have worked for more than a decade to make efficiency a usable idea in the social service enterprise. Not every conceptual hurdle we came upon has been overcome, but we have yet to find one that we have been forced to give up on. In a word, we know that efficiency "works." Our single greatest disappointment has been our inability to find more than the smallest handful who share this enthusiasm.

REFERENCES

Aigner, S.M., & Simons, R.L. (1977). Social work and economics: Strange bedfellows. *Social Work*, *22*, 305-308.

Burns, E. (1974). Some economic aspects of welfare as an institution. In J.M. Romanyshyn (Ed.), *Social science and social welfare*. New York: Council on Social Work Education.

Carter, R. (1983). *The accountable agency*. Beverly Hills: Sage Publications.

Culyer, A.J. (1973). *The economics of social policy*. London: Martin Robinson.

Davies, B., & Knapp, M. (1988). Introduction: The production of welfare approach; Some new PSSRU argument and results. *British Journal of Social Work*, *18*, 1-11.

Elkin, R. (1985). Paying the piper and calling the tune: Accountability in the human services. In S. Slavin (Ed.), *Managing finances, personnel, and information in human services*. New York: The Haworth Press.

Gates, B. (1980). *Social program administration: The implementation of social policy*. Englewood Cliffs, NJ: Prentice-Hall, Inc.

Giannetti, V., & Wells, R. (1985). Professional outcome and professional affiliation. *Social Service Review*, *59*, 32-43.

Gilbert, N. (1983). *Capitalism and the welfare state: Dilemmas of social benevolence*. New Haven, CT: Yale University Press.

Gruber, M. (1986). A three-factor model of administrative effectiveness. *Administration in Social Work*, *10*(3), 1-14.

Hanrahan, P., & Reid, W.J. (1984). Choosing effective interventions. *Social Service Review*, *58*, 244-258.

Hasenfeld, Y. (1984). The changing context of human services administration. *Social Work*, *29*, 522-529.

Judge, K., & Knapp, M. (1985). Efficiency in the production of welfare: The public and the private sectors compared. In R. Klein & M. O'Higgins (Eds.), *The future of welfare*. Oxford: Blackwell.

Johnson, W., & L'Esperance, J. (1984). Predicting the recurrence of child abuse. *Social Work Research and Abstracts*, *20*, 19-25.

Kanter, R., & Summers, D. (1987). Doing well while doing good: Dilemmas of performance measurement in nonprofit organizations and the need for a multiple-constituency approach. In W.W. Powell (Ed.), *The nonprofit sector*. New Haven: Yale University Press.

Kemper, P., Applebaum, R., & Harrigan, M. (1987). Community care demonstrations: What have we learned? *Health Care Financing Review*, *8*, 87-100.

Miller, L. (1988). Increasing efficiency in community-based long-term care for the frail elderly. *Social Work Research and Abstracts*, *24*, 7-14.

Miller, L., & Pruger, R. (1988). Efficiency and fairness in the practice of care. Unpublished manuscript.

Miller, L., & Pruger, R. (1989). The two activities of social services: Maintenance and people changing. In R.T. Golembiewski et al. (Eds.), *Public budgeting and financial management*. New York: Dekker.

Neinberger, J., & Wildavsky, A. (1973). *The budgeting and evaluation of federal recreation programs*. New York: Basic Books.

Neugeboren, B. (1985). *Organization, policy, and practice in the human services*. New York: Longman.

Newman, E., & Turem, J. (1985). The crisis of accountability. In S. Slavin (Ed.), *Managing finances, personnel, and information in human services*. New York: The Haworth Press.

Page, A. (1977). Economics and social work: A neglected relationship. *Social Work*, *22*, 48-53.

Patti, R. (1987). Managing for service effectiveness in social welfare: Toward a performance model. *Administration in Social Work*, *11*(3/4), 7-21.

Pruger, R. (1987). *Making information work: The story of the equity project*. Unpublished manuscript.

Rapp, C., & Poertner, J. (1987). Moving clients center stage through the use of client outcomes. *Administration in Social Work*, *11*(3/4), 23-38.

Reid, W.J. (1987). Service effectiveness and the social agency. *Administration in Social Work*, *11*(3/4), 41-58.

Reid, W.J., & Hanrahan, P. (1982). Recent evaluations of social work: Grounds for optimism. *Social Work*, *27*, 328-340.

Reisch, M., & Taylor, C. (1985). Ethical guidelines for cutback management. In S. Slavin (Ed.), *Managing finances, personnel, and information in the human services*. New York: The Haworth Press.

Rhoads, S. (1985). *The economist's view of the world: Government, markets, and public policy*. Cambridge: Cambridge University Press.

Rubin, A. (1985). Practice effectiveness: More grounds for optimism. *Social Work*, *30*, 469-476.

Simon, H. (1957). *Administrative behavior*. New York: The Free Press.

Thomlinson, R.J. (1984). Something works: Evidence from practice effectiveness studies. *Social Work*, *29*, 51-56.

Tucker, M. (1974). Interfacing economics with social work education. *Journal of Education for Social Work*, *10*, 96-104.
Weissert, W. et al. (1986). The past and future of home- and community-based long-term care. *The Milbank Quarterly*, *66*, 309-388.
Williams, A., & Anderson, R. (1975). *Efficiency in the social services*. Oxford: Basil Blackwell.

PART 2: ON COSTS, BENEFITS, AND EFFECTS

Cost

Martin Knapp, PhD

EFFICIENCY, AUSTERITY, AND ECONOMICS

To the beleaguered social work practitioner, bombarded with needs but no longer showered with resources, the era of efficiency is like a bad dream. The pursuit of efficiency appears as a thinly veiled euphemism for "cut," and an attack on the fundamental caring principles of social work. Exposed to explanations that efficiency is not all bad – that it is concerned with getting more services and having greater impact from the same pot of money, or doing as well as before but using fewer resources – the downtrodden social worker can see only the word "cost." Cost and care simply do not look like comfortable bedfellows. Policies directed at the management of one seem to threaten practices dominated by concerns with

---

Dr. Knapp is Professor of Economics of Social Care, and Deputy Director, Personal Social Services Research Unit, University of Kent at Canterbury, Canterbury, CT2 7NF, England.

This paper was written while the author was International Fellow in the Institute for Policy Studies, Johns Hopkins University. It is based on research mainly undertaken in the PSSRU and substantially funded by the UK Department of Health and the Economic and Social Research Council.

the other. Unfortunately, the cost information that needs to go into almost any conceptualization of efficiency has often been hastily requested, poorly constructed and naively employed. Cost, not surprisingly, has a bad press.

That image of cost can be altered by looking at how efficiency inquiries should ideally be conducted—one of the purposes of this volume—and by drawing lessons from economics, the discipline which has devoted most attention to the topic. From the study of scarcity, economics has both accumulated theoretical experience and developed comparative empirical advantage in the design and execution of cost research. Some of those theoretical perspectives have already been brought into the discussion of social work, and some have been used to develop a conceptual framework within which to locate empirical investigations of efficiency (Knapp, 1984).

Thankfully, the arrival of the economist is not seen in the same light by everyone—liberating knight in shining armor to some, vanguard of a hated army of occupation to some others—and in neither fantasy guarantees a resolution of resource allocation problems. Economists are as fallible as anyone else, and there is the additional problem that their involvement is commonly too little, too late, or too remote. What is needed is the premeditated and purposive integration of costs with other dimensions in the policy or practice debates; and this means the examination of hypotheses which are not "cost questions" or, say, "outcome questions," but are the types of question faced by the policy-maker who has to balance resources against achievements, efficiency against equity, irate social work managers against persistent accountants.

FOUR COST-RULES

There are four simple rules to observe in looking at the costs of social care services.

1. Costs should be measured comprehensively.
2. The cost variations that are inevitably revealed in any empirical exercise—variations between clients, facilities, areas of the country, and so on—offer a wealth of information which should be explored and exploited.

3. Variations encourage comparisons—this service is cheaper than that one, this group of clients is more costly to accommodate than that group—and only like-with-like comparisons have full validity.
4. Finally, cost information should be integrated with information on client outcomes. (It is as wrong to examine costs in the absence of outcome information as it is to study outcomes without paying heed to costs.)

These four rules—which, of course, are easier to recite than to observe in practice—provide the framework for the remainder of this paper. The following section looks at comprehensive costs, with illustrations from the long-term care of people with learning difficulties (mental handicaps). Other sections examine cost variations and like-with-like comparisons, focusing on residential care of elderly people and foster family care for children. A final section considers the integration of costs and outcomes, with evidence from the community care of people with mental health problems. Recent British research is used for the illustrations.

COMPREHENSIVE COSTS

Why Go Comprehensive?

Most clients receive more than one service, probably from more than one agency. The greater and the more diverse a client's needs, the broader the range of services likely to be utilized. Any evaluation which addresses the resource dimension should therefore aim to cost every service within a care "package." Even if the aim of an evaluation is simply to compare the costs of running, say, local authority and non-profit hostels for people with learning difficulties, it would be dangerous to define costs narrowly to exclude other services received by residents, for combinations of services tend not to be thrown together willy-nilly. The staff of a local authority hostel may "network" in different ways from the staff of a non-profit hostel; e.g., perhaps better links with the health service but less access to day care provision, more opportunity to help clients obtain social security entitlements, but less independence to advocate for change. The purposive combination of services to meet needs is one of the underlying principles of case management, and in Britain at least, this mode of working—preferably with budgeting decisions devolved down to case managers—will soon become the norm (Cm 849, 1989). It is hard to imagine effective case management in an age of relative austerity proceeding in a cost vacuum.

There is another compelling reason for comprehensiveness. In Britain, most long-stay hospitals for people with learning difficulties (and for other client groups) are scheduled to run down to closure (gradually) over the next decade. Savings from the rundown of hospitals are expected to fund the running costs of the community services scheduled to replace them. Cost comparisons between hospital and community services are essential in order to plan or monitor these financial transfers. To make sensible comparisons, and to ensure that new service developments are adequately funded, cost measures need to be equally comprehensive.

The comprehensive costing of social work services raises a number of conceptual and practical issues. It is also controversial. It has been pointed out, for example, that it is iniquitous to include daily living expenses, housing rents and routine visits to the family practitioner in the costs of community care, since, so the argument runs, every one of us has these basic needs and incurs equivalent costs. But only if utilizations of these services is identical for the options being studied is it legitimate to ignore the associated costs.

How Comprehensive?

There must obviously be practical limit to any empirical exercise. An evaluation of long-term care cannot go on for as long as clients receive services before reaching some conclusion about costs (or effects). So, even though many practice decisions are deliberately intended to have lifespan effects, and even though the many policy implications outlive the clients immediately affected by them, costs, like outcomes, tend only to be measured over relatively short intervals. The practical question is to decide when short is "too short," and to avoid drawing conclusions from inadequate evidence. The perpendicular question concerns the services, activities and expenditures to include. Should we, for example, include the forgone earnings (and associated losses in national productivity) of people with learning difficulties? Should we include the burden of informal care falling to relatives or neighbors? Is a social security benefit a cost? It will probably come as no surprise to learn that there is no simple yes/no answer to each of these questions, for the answer depends on the context and on the policy questions being addressed. But the questions do need to be asked.

How Are the Costs Measured?

Routinely quoted cost figures come from agency accounts. They are either total cost (actual spending on a service, usually during one financial year, perhaps with the addition of a notional capital element) or average cost (this amount divided by some measure of workload, such as the number of clients seen or the number of "resident days"). What is available is not the same as theory would recommend for employment in empirical research. Ideally, we need marginal opportunity cost, whereby marginal is meant the addition to total cost attributable to the inclusion of one more client or the delivery of one more unit of service, and opportunity indicates reference to opportunities forgone. The immediate cost of "slotting" one more person into a day program or accommodating one more resident in a hostel will be small. But, if policy intentions are, to substitute community services for all or most long-term hospital beds for people with learning difficulties, it makes no sense to use this smaller short-term measure since there is an obvious limit to the number of people who can be "slotted" into extant services. It happens, however, that today's (short-run) average revenue cost, plus appropriately measured capital and overhead elements, is probably close to the long-run marginal cost.

Opportunity costs measure the true private or social value of resources, their values in alternative uses. In a perfectly informed and frictionless market economy, opportunity cost would be identical to market price. But not everything is marketed, not every market works smoothly, and information is rarely complete, with the result that observed prices and opportunity costs diverge. Thus, the recorded depreciation payments on capital equipment or buildings will not usually reflect the opportunity costs of using these durable resources, and an estimate of replacement cost or sale value will be closer to the underlying theoretical concept. Nor will the (zero) payments to volunteers necessarily indicate their social value. The principle of opportunity costing should never be abandoned, but enthusiasm for it will have to be tempered when it comes to application.

An Example of Comprehensive Costing

In 1983, the UK Department of Health and Social Security announced the Care in the Community demonstration program to promote and evaluate the development of community care services for people who had lived in hospital for many years, but who probably no longer needed to be there. It was believed that community living could be better for them, but there were few community facilities to receive them, and little research to confirm or deny the belief. The program covered all long-term care groups, including people with learning difficulties, and set out to rehabilitate about 900 people into a wide variety of community care arrangements (Renshaw et al., 1988). The care processes, client outcomes and costs of the program were evaluated by the PSSRU (Knapp et al., 1990).

Hospital and community costs were examined. The former were based on average revenue expenditures for whole establishments, disaggregated down to the ward level, plus a capital element based on an annuitized valuation of site and durable resources, and an element for service inputs from outside the hospital budget, such as social work, education or volunteer visiting. The costs of the "packages" of community care were calculated from information on service utilization gathered in interviews with staff (often case managers) nine months after each client was discharged from the hospital. These are as close to long-run marginal (opportunity) costs as data would allow. Social security payments were used as proxies for living expenses (a common assumption) and were of direct interest in the examination of the distributional consequences and funding of community care.

Comprehensive community care costs, arranged by the type of accommodation, are summarized in Table 1. The largest single component of cost is accommodation, which often includes much of the staff support needed by clients, but "community care" does not stop with the place of residence. It is not necessary to dwell on

Table 1: Comprehensive Community Care Costs, by Place of Residence, for People with Learning Difficulties (Mental Handicaps) formerly Long-Stay Hospital In-Patients

Community Care Component	Percentage of Total Package Cost by Place of Residence in the Community[1] RESID %	HOST %	SGH %	SHELT %	USGH %	INDEP %	FOST %	All Places %
Accommodation and living expenses [2]	62	75	58	50	33	36	42	63
Project overheads [3]	26	8	22	29	33	34	31	18
General practitioner	1	0	0	1	1	0	0	0
Day activity services	2	10	4	5	4	8	12	7
Education	2	5	8	3	0	7	7	6
In-patient hospital	1	0	1	0	5	0	0	1
Out- and day patient hospital	0	0	0	1	1	0	0	0
Community health care [4]	2	0	3	2	1	0	4	2
Community social care	4	1	3	10	22	14	3	3
Other [5]	0	0	0	0	0	0	1	0
(%)	100.0	100.0	100.0	100.0	100.0	100.0	100.0	100.0
(N)	6	86	67	14	6	6	9	194

Notes:

1. Accommodation codes are as follows: RESID = residential home; HOST = hostel; SGH = staffed group home; SHELT = sheltered housing; USGH = unstaffed group home; INDEP = independent accommodation; FOST = adult fostering.
2. Includes staff costs where staff are employed solely at place of residence and included in accommodation budget.
3. Inclusions vary from project to project, usually covering some immediate managerial support and some peripatetic staff support.
4. Includes community nursing, medical practitioners not attached to hospital or family practice, miscellaneous community health paramedic services (chiropody, dietary services, incontinence service, ...).
5. Includes police, travel not covered elsewhere and miscellaneous minor services.

the detailed results, but we should note three things. Comprehensive costing is needed to gauge the respective implications of community care for public, voluntary and private agencies and for the personal resources of clients. Second, consistency of definition is needed when making comparisons with hospital since most of community services have already been subsumed within hospital budgets. Third, comprehensive costing is feasible.

EXPLORING AND EXPLOITING VARIATIONS

Inter-client and inter-facility variations in cost are generally marked. They should not be ignored. Good social work practice is individually tailored, and practitioners talk about the "average client" only in order to make some temporary general remarks. Service packages reflect individual needs, and because needs vary, so too should costs. The correspondence is not perfect, but is sufficient for us to have reservations about those policy recommendations based solely on averages. Unless the organization of care is so routinized or regimented as to disregard utterly individual circumstances, the costs of care will reflect client differences. Service responses are also shaped by, *inter alia*, the preferences and perspectives of other professionals working with these clients, the bureaucratic tendencies of organizations, their scales of operation, and the characteristics of local economies and their implications for the price of labor.

The "Behavioral" Cost Function

A large and impressive body of applied economics research has explored the extent of, and reasons for, cost variations. The most favored route is to estimate a statistical cost function – a multivariate technique designed to simultaneously tease out the many influences on cost. The cost function has impeccable theoretical credentials – it follows directly from the economic theory of production and is based on assumptions which have been exhaustively debated in a wide variety of market and non-market contexts – and in application has proved manageable and informative (Knapp, 1984). If it is reasonable to make the analogy between, on the one hand, the production model of inputs and outputs that is the bread and butter of microeconomics and, on the other, the processes of delivering social work services, then the cost function has currency. It is, however, a data-hungry tool, and in the hands of the inexperienced, uninitiated or unscrupulous it is as dangerous and misleading as any other empirical procedure.

The cost function is most commonly estimated for a cross-section of "production units" known or assumed to have similar objectives and to employ reasonably similar production techniques. In the study of manufacturing goods, the study of costs would ideally be based on data from a sample of factories (or work teams within factories) producing a similar if not identical product. In human service applications, "production" is certainly not routinized or standardized, and the terminology of business is unfamiliar, but the production metaphor has considerable validity (Davies & Knapp, 1988a). Using economic theories and analogies, the "production of welfare" approach can organize large parts of the social work, philosophy, psychology, and sociology literatures on social care. It can do so in ways which are most likely to maximize the applicability of economic tools of analysis (of which the cost function is one), but least likely to jettison either the principles of social work or the empirical insights that come from other disciplines. Under the terms of this production of welfare approach the cost function has a behavioral interpretation.

In practice, a cost function is a multiple regression equation, with dependent variable either total or average cost, and the explanatory or independent variables suggested by a pragmatic blend of theory and experience. These determinants of cost are selected in an actual empirical study on these same criteria plus statistical significance.

Cost Variations Between Residential Care Homes

The cost function can be illustrated by looking at residential care homes for elderly people. For simplicity, consider only those homes in the public sector. Even when aggregated up to the local authority level, average revenue cost per resident week varies by a factor of more than two, and within these areas the variations are even greater. Yet, these homes all concentrate mainly on delivering long-term care for what one would expect to be similar populations of frail elderly people, hiring staff covered by nationally agreed terms of employment and salary scales, governed by common regulations, enmeshed within similar care systems, and collectively guided by more than thirty years of research-based social work training. The cost variations might therefore reflect slight differences in these factors as well as the efficiency with which facilities deliver residential care services of a given quality and impact upon the well-being of residents.

A cost function was estimated using data for a nationally representative sample of homes in England and Wales. Average revenue cost per resident place normally in use was taken as the dependent variable. (Capital costs and any costs falling outside homes' budgets were omitted.) The variables examined as possible (statistical) explanations for the cost differences covered resident characteristics (especially dependency or activities of daily living), staff, homes, non-residential services, and the localities and labor markets within which homes were located. The investigation was underpinned by theoretical arguments (Davies & Knapp, 1981), and informed by previous empirical research. After a deal of statistical analysis, the "best" cost function most closely met the conventional criteria of parsimony, statistical significance and interpretability in the light of theoretical arguments (Table 2). It is unnecessary fully to detail the function here, but some of its policy conclusions illustrate both the power of the technique and the need to explore cost variations (Darton & Knapp, 1984).

The cost function "explained" 76% of the observed variation in average cost; that is, the differences between homes were largely accounted for by the factors listed above. This emphasizes the need to examine costs in their proper context—indeed, some high cost homes and local authorities proved to be costing less than would have been expected once resident dependency and home, labor market and other factors had been taken into account. The residual or unexplained 24% of the variation in cost might then be interpreted as picking up differences in efficiency, although there is a need for caution here. The cost function also tests for the existence of economies and diseconomies of scale. In this case average cost per resi-

Table 2: An Estimated Average Operating Cost Function for Old People's Homes, 1981-82

The estimated cost function indicates that average expenditure on manpower and running expenses per bed normally in use per resident week is equal to:

−552.826*		
+1.743**	×	labour cost index
−3.264**	×	no. of beds normally in use in group-living homes
−1.604**	×	no. of beds normally in use in non-group homes
+0.024*	×	no. of beds normally in use in group-living homes, squared
+0.014**	×	no. of beds normally in use in non-group homes, squared
+10.725*	×	percentage occupancy level (at 31/10/81)
−0.058*	×	percentage occupancy level (at 31/10/81), squared
+5.995*		if staff of home have duties in sheltered housing
+4.358*		if all laundry is done within the home
+7.736*	×	no. of permanent admissions in 12 months to 31/10/81 per no. of beds normally in use
+3.188*	×	no. of short-stay discharges in 12 months to 31/10/81 per no. of beds normally in use
+4.927*		if mixed-sex non-group-living home
+34.823**		if single-sex group-living home
+63.294**		if mixed-sex group-living home
−12.488**	×	proportion of residents in limited dependency category
+51.714**	×	proportion of residents in appreciable or heavy dependency categories
−0.814**	×	no. of beds normally in use weighted by proportion of residents in appreciable or heavy dependency categories

Significance levels: * $0.01 \leq p < 0.05$; ** $p < 0.01$
F – value for equation; $F = 37.58$, $p < 0.0001$
$R^2 = 0.76$, Adjusted $R^2 = 0.74$
$n = 218$

dent was found to be lowest at a scale of between 55 and 70 beds (depending on whether or not the home was designed around group living units). This scale is larger than that preferred on professional grounds as "noninstitutional."

Not surprisingly, resident dependency and cost are positively related. The advantage of the cost function is that it quantifies the cost-dependency association and can therefore feed into debates about the balance of care between, say, hospital, residential and domiciliary settings (Davies & Knapp, 1988b). Residential homes in the UK also sometimes act as day facilities for nonresidents, or the hub of a meals-on-wheels service, and via an empirical inquiry of this kind we can separate the joint costs of these different services.

The cost function approach offers other insights and is equally relevant in other social work contexts. Two general points must be made. First, social care and, therefore, social care costs are inherently variable. Second, the variations tell us a great deal about practice and policy. We should not shy away from them.

LIKE-WITH-LIKE COMPARISONS

Cost variations invite comparisons. Why is one residential home more costly than another? Why does one group of residents impose a greater care burden than another group? Why is residential care more expensive in some parts of the country than others? In answering these important questions, the problem is to avoid drawing inferences from misplaced or misjudged comparisons. Only like-with-like comparisons offer admissable evidence.

Consider the comparative costs of foster family care and residential homes for children who cannot be accommodated with their parents (Knapp & Fenyo, 1989a, 1989b). Foster family care has long been preferred to residential care for the majority of children who cannot be accommodated with their natural parent(s). This preference has been couched overwhelmingly in terms of child welfare outcomes, and does not apply to every child in care (Parker, 1988). If we put these "welfare" arguments to one side for the moment—they are not to be ignored, of course—we can examine the often made claim that foster family care is cheaper than residential accommodation, a claim that is frequently used to bolster the "welfare" arguments.

The two placement types do not accommodate identical populations of children. Those in residential care are typically more "difficult" than children boarded out—they are more likely to have physical and mental handicaps, delinquent tendencies or backgrounds, and emotional or behavioral problems such as aggression, hyperactivity or propensity to self-mutilation (Millham et al., 1986;

Packman, 1986). The relevance of these differences between the two child care populations stems from the cost correlates of "difficulty." Evidence of cost-difficulty associations comes from cost function explorations of the kind described earlier (Knapp & Fenyo, 1989a; Knapp & Smith, 1985; Young & Finch, 1977). For foster family care, for example, the weekly boarding out payment and the costs of social worker time were found to be associated with: child characteristics (such as age, gender, legal status, health status, emotional and behavioral characteristics, educational needs, previous placement and time in care); placement experiences (duration to date, location, composition and previous fostering experience of the family, objectives and plans, housing size and tenure); characteristics of the natural family (parents alive, visiting frequency, their own long-term care experience, siblings); and some features of the supervising social worker.

The differences between the two child care populations, and the links between child characteristics and costs, have important implications for the current policy emphasis in Britain on shifting the balance of care away from residential accommodation. In projecting the savings to be reaped from shifting the balance towards foster family care, care must be taken only to compare like with like. It is, for example, all too easy to exaggerate the cost differences between placement options. Simple cost averages for residential and foster family care are compared without paying heed to the child characteristic differences that lie behind them. If a child whose need-generating characteristics are typical of the residential population moves from such a setting to a family placement, the costs of the latter will be somewhat higher than today's foster care average. If *today's* average costs are employed to justify savings, and to justify a changing balance of care, there is a danger of grossly underpitched funding for family placements.

There are two corollaries. The most common way to alter the balance of care is to place with families children whose needs are more difficult or demanding to meet than the average for the foster care group but less difficult than the average residential child. The effect will be to raise the average difficulty and average cost in both foster and residential care. This is seemingly a perverse consequence of a policy which has cost savings as one of its aims, but in fact the total costs of the child care system will fall. This helps to explain the (constant-price) cost inflation in both services in England in the last few years. The proportion of children boarded out has risen from 33% in 1976/77 to more than 50% today, and over the same period the costs of both placement types have inflated by more than 30%. Further changes in the balance of care will mean higher direct and hidden costs for foster care. But residential care will then also be accommodating a more difficult clientele, raising its average cost, and making it more conspicuous to politicians and policy-makers in budget-conscious agencies.

The second corollary concerns placement breakdowns. There is evidence to suggest that local authorities placing a relatively low proportion of children in their care with families have higher rates of successful placement endings (fewer breakdowns) in both the foster family and residential home populations (Rowe, 1989). This exactly parallels the cost implications. Placement breakdown is by no means the only indicator of outcome, but higher breakdown rates could be expected from the under-resourcing of services, since boarding out payments to families would then be lower, and there would be less support for families from social workers.

In common with a number of today's debates about social care, the encouragement of foster family care as an alternative to residential care is based primarily on assumptions (and some evidence) about comparative effectiveness gauged in terms of client welfare, and reinforced by assumptions (but usually *no* evidence) about comparative costs. But valid comparisons are made all to infrequently, with the result that major resource implications of care decisions are overlooked, wide variations around averages remain unexplained (and unexploited) and the system effects of changes in the balance of provision are missed.

The research evidence that has been gathered to date for the two principal forms of out-of-home child care may not be perfect, but it has cumulative force. It does not imply that recommendations to expand the proportion of children in care in family placements rather than residential settings are inappropriate. It implies that the commonly employed cost argument for this expansion seriously underestimates the future costs of foster family care and dangerously exaggerates the savings that will flow from changes in the balance of care.

MERGING COSTS AND OUTCOMES

It is dangerous to separate discussions of costs and outcomes. Ideally, the two types of information will be merged in any policy review, research endeavor or practice decision. The pragmatic second best is to recognize that one source of information is missing when drawing conclusions from the other. Costs and outcomes are two parts of the same production system—the inputs and outputs, means and ends, causes and effects.

The Care in the Community demonstration program, introduced earlier, sought to place long-stay hospital residents in well-supported community settings. Among those served by the program were 226 people with mental health problem. A comprehensive costing of community care revealed wide variations, but also showed that community care was less expensive (on average) than hospital. This prompts two questions. Is community care being provided "on the cheap" in that clients are left with a lower quality of life they enjoyed in the hospital? Second, were more expensive care programs also the more successful in improving or maintaining the welfare of clients?

The Outcomes of Dehospitalization

The client outcome findings for the program were encouraging (Knapp et al., 1990). All clients enjoyed a satisfactory standard of care and support in the community, and none of the often posited failures of dehospitalization policies occurred. Nobody was "dumped" to drift without support, nobody was destitute or homeless, and nobody was imprisoned. There were few readmissions to hospital, and most were of short duration and for reasons unconnected to mental ill health. Mortality rates were low, and the suicide rate was less than 1% in the first nine months after discharge from the hospital. The burden on informal carers was very small, for the sad reason that few clients had informal care settings to which they could move. (The median length of continuous stay in the hospital prior to the move was over twelve years and most clients had either lost contact with relatives, or had relatives who were unable to make significant contributions to their care.) No pervasive evidence of unreasonable adverse effects on neighborhoods could be found.

These "standard of living" outcomes are important, but the evaluation was more concerned with changes in client welfare assessed along various dimensions. Here, too, the community care arrangements proved successful. Integration into "ordinary" community life-styles was not perfect but far better than in the hospital. Twice as many people expressed positive attitudes about activities in the community compared to their preferences in the hospital. Most had more choice about how they spent their time, although they were constrained by social security benefits and few "real" employment opportunities. Reported social contacts were more numerous than in the hospital. Skill levels, behavioral problems and symptoms (as reported by staff) showed no differences between hospital and community. Self-report measures of satisfaction with the environment, psychosocial functioning, depression, satisfaction with social interaction, and global morale revealed borderline improvements after the move from the hospital.

Thus the statistical testing of changes in client welfare over the period from hospital residence to a community setting nine months after discharge gives no reason to believe that community living is going wrong—negative outcomes had been avoided—and some suggestions that client quality of life was better than in the hospital. The lower costs of community care were not the result of poor quality care. Further interpretative research lead to the conclusion that the case management arrangements put in place by the demonstration program were responsible for these and other favorable outcomes. This finding is consistent with work in related areas (Davies & Challis, 1986; Challis et al., 1989), and has had a marked impact on subsequent proposed legislation (Cm 849, 1989).

Costs and Client Outcomes

Are the costs of community care packages associated with differences in levels of need and/or changes in need (outcomes)? With the accumulation of experience on the needs and preferences of mentally ill people living outside the hospital, with the growing emphasis on efficiency in resource utilization, and with the growth of case management and similar procedures, there are good reasons for expecting strong associations between costs, client characteristics and outcomes. These associations were examined by estimating a series of cost functions. Using the scores on a number of quantitative instruments in the hospital and in the community gave us both "absolute" and "relative" (change) measures of client welfare or needs. These client-specific measures were used as predictors of cost differences in the cost function analyses, along with measures of various mediating factors. The detailed results are not reported here, but some interesting associations were uncovered (Knapp et al., 1990).

Community care costs were found to be sensitive to most of the client characteristics for which measures were developed. Costs were higher for people displaying greater needs along the dimensions spanned by the scales for skills, behavioral problems and symptoms, social interaction, life satisfaction, and personal presentation. Costs were higher when client needs were greater, providing an indication of the responsiveness of resources to the health and social welfare needs of clients. The consistency of response was again largely attributable to case management practices.

In contrast, changes in client characteristics played only a small role in the explanation of cost variations, even though statistical significance was achieved after standardizing for differences in client background circumstances, hospital-based assessments of need, and some organizational features of the different demonstration projects. Higher costs were associated with the attainment of greater or better outcomes in relation to behavioral problems, symptoms, morale, personal presentation, and depression. It is interesting to note, therefore, that although client welfare improvements were modest at an aggregate level, once we delve into inter-client variations we uncover significant positive associations with cost.

Following the fourth costing rule, therefore, offers evidence that higher community care costs are attributable to both higher levels of observed need and greater achievements in reducing this need. Higher levels of spending on community care will generate further improvements in quality of life. Whether the additional expenditure is warranted by the welfare improvements cannot be examined here.

CONCLUSION

Cost analyses will or should never "make decisions," but they can "make decisions better informed" (Weisbrod, 1979). Cost information is an indispensable part of an efficiency analysis, and efficiency considerations should inform almost all policy and practice discussions in areas like social work which embody such important social perspectives and which often are so dependent on public subsidies. Consequently, policy or practice changes which ignore costs, or which embody cost information without obeying or recognizing the four basic rules discussed here, will usually be of dubious validity, and will sometimes offer advice which is dangerously inaccurate.

This is inexcusable, for, as the illustrative studies in this paper have shown, cost analysis need not be an horrendous, ideologically compromising or scientifically complex task. There are enough examples of bad costs research in the social work literature to demonstrate that it is not as simple as some people may think, but there are also enough examples of good research to encourage further attempts.

REFERENCES

Cm 849 (1989). *Caring for people: Community care in the next decade and beyond*. London: Her Majesty's Stationery Office.

Challis, D.J., Darton, R.A., Johnston, L., Stone, M., Traske, K., & Wall, B. (1989). *Supporting frail elderly people at home*. Canterbury: Personal Social Services Research Unit.

Darton, R.A., & Knapp, M.R.J. (1984). The cost of residential care for the elderly: The effects of dependency, design and social environment. *Aging and Society*, *4*, 157-183.

Davies, B.P., & Challis, D.J. (1986). *Matching resources to needs in community care*. Aldershot: Gower.

Davies, B.P., & Knapp, M.R.J. (1981). *Old people's homes and the production of welfare*. London: Routledge and Kegan Paul.

Davies, B.P., & Knapp, M.R.J. (Eds.). (1988a). The production of welfare approach: Evidence and argument from the PSSRU. *British Journal of Social Work, 18* (whole supplement).

Davies, B.P., & Knapp, M.R.J. (1988b). Costs and residential social care. In I. Sinclair (Ed.), *Residential care: The research reviewed*. Volume 2 of the Wagner Report. London: Her Majesty's Stationery Office.

Knapp, M.R.J. (1984). *The economics of social care*. London: Macmillan.

Knapp, M.R.J., Cambridge, P., Thomason, C., Beecham, J., Allen, C., & Darton, R.A. (1990). *Care in the community: Evaluating a demonstration programme*. Aldershot: Gower.

Knapp, M.R.J., & Fenyo, A.J. (1989a). Economic perspectives on foster care. In P. Carter, T. Jeffs, & M. Smith (Eds.), *Social Work and Social Welfare Yearbook 1989*.

Knapp, M.R.J., & Fenyo, A.J. (1989b). Efficiency in foster family care: Proceeding with caution. In J. Hudson & B. Galloway (Eds.), *State intervention on behalf of children and youth*. Dordrecht: Kluwer.

Knapp, M.R.J., & Smith, J. (1985). The costs of residential child care: Explaining variations in the public sector. *Policy and Politics, 13*, 127-154.

Millham, S., Bullock, R., Hosie, K., & Haak, M. (1986). *Lost in care*. Aldershot: Gower.

Packman, J. (1986). *Who needs care*? Oxford: Basil Blackwell.

Parker, R.A. (1988). Residential care for children. In I. Sinclair (Ed.), *Residential care: The research reviewed*. Volume 2 of the Wagner Report. London: Her Majesty's Stationery Office.

Renshaw, J., Hampson, R., Thomason, C., Darton, R.A., Judge, K.F., & Knapp, M.R.J. (1988). *Care in the community: The first steps*. Aldershot: Gower.

Rowe, J. (1989). Paper presented at NATO conference, Maratea.

Weisbrod, B.A. (1979). A guide to benefit cost analysis, as seen through a controlled experiment in treating the mentally ill. Unpublished paper. Institute for Research on Poverty, University of Wisconsin, Madison.

Young, D.R., & Finch, S.J. (1977). *Foster care and nonprofit agencies*. Lexington: D.C. Heath.

Methodological Issues in Efficiency-Focused Program Evaluation: The Case of Mental Health

Sid Frankel, MSW

One of the major components of evaluation research involves the process of assigning relative value as a means for distinguishing good from bad programs (Cook & Shadish, 1986; Scriven, 1980). The cost-benefit framework provides one comprehensive means of discriminating among program alternatives through determining which utilizes resources in the most efficient manner. In more pragmatic terms, this approach to valuation can be seen as providing the means to address three equivalent evaluation questions: (1) Which alternative provides the most output for the least expenditure of resources? (2) Which alternative provides a given level of output for the least cost? (3) Which output provides the most output for a given level of input?

This paper focuses upon methodological issues in implementing three activities which are unique to this approach to program evaluation. These are: (1) analysis of costs, (2) analysis of benefits, and (3) discounting of costs and benefits. Examples from the mental health field are utilized to explicate various means of dealing with the methodological challenges related to these activities.

Mr. Frankel is Assistant Professor and Director of the Child and Family Services Research Group, Faculty of Social Work, University of Manitoba, Room 418B Tier Building, Winnipeg, Manitoba, Canada, R3T 2N2, and Senior Analyst, Program Development and Support Unit, Day Care, Youth, and Employment Support Division, Department of Family Services, Manitoba.

Research for this study was supported in part by a National Welfare Fellowship from the Government of Canada and educational leave funding from the Government of Manitoba.

ANALYSIS OF COSTS

Cost-benefit and cost-effectiveness approaches to program evaluation generally should involve an experimental or quasi-experimental design which is utilized to determine which costs and benefits can actually be attributed to the implementation of a particular alternative. Weisbrod and Schlesinger (1981) have argued that this is as important for costs as for benefits, utilizing the example of the need to discriminate between the food costs attributable to a residential program and the value of the food which the participants would consume even if they did not attend the program.

Costs can be categorized as direct or indirect, with the former referring to the value of all resources expended in actually producing the service, and the latter referring to the value of resources which are not identified as inputs in the production function, but which must be withdrawn from other productive activity in order for the service to be delivered and consumed (Frank, 1981). The time of the client and the client's family related to receiving the service are examples of indirect costs.

The analysis of costs and benefits can be conducted from the perspective of program providers or funders, clients, or society at large (Rossi & Freeman, 1985). Cost assessment from the program funder's perspective would include only those direct costs borne by the funder; while assessment from the client's perspective would include the portion of both direct and indirect costs borne by the client. A social perspective would attempt to account for all resources directly and indirectly utilized in the production and consumption of the service while excluding transfer payments (Mishan, 1972). The social perspective, as well as that of government as funder, is usually required in cost-benefit and cost-effectiveness analyses conducted in support of public investment decision-making. Furthermore, unless costs borne by all constituencies are measured comprehensively, any cost difference found among alternatives may be due to shifts of costs into unmeasured forms rather than to differences in actual social cost (Sharfstein & Clark, 1978).

Costs are conceptualized according to an economic or opportunity doctrine:

> The economic cost of any resource is the payment required to keep the resource in its present employment. Equivalently, the economic cost of any resource is the remuneration that the resource would receive in its best alternative employment opportunity. (Nicholson, 1982, p. 230)

Thus, from a social perspective, the assessment of the costs of each alternative should include the value of all direct and indirect resources required for its implementation. This value is determined according to the best alternate use of the resource which was foregone.

The assessment of costs involves three interrelated processes: (1) the identification of resources, (2) their measurement, and (3) their valuation. The last two are synonymous where market prices exist and accurately reflect social values.

Identification

The identification of costs from a social perspective begins with a broad accounting of all resources utilized in the production, delivery, and consumption of the service. Levin (1983a; 1983b) has developed the ingredients approach for determining the costs of the program being assessed. It involves the systematic listing of resources according to the following classes: personnel, facilities, materials and equipment, client time and other client inputs, and a residual category.

This approach could also be utilized to identify the indirect costs of health, social, and public protection services provided and financed by parties external to the organization providing the direct treatment service being evaluated. Such indirect treatment costs often constitute a significant proportion of total cost for all but the most debilitated institutionalized clients in this age of deinstitutionalization (Cannon, 1985). For example, in Weisbrod's (1983) study, described above, indirect treatment costs constituted 35% of total costs for the inpatient control group, and 27% of total costs for the experimental group served by the community-based service. Similarly, the identification of costs should include indirect resources contributed by the client, his family and collaterals. Material and psychological resources contributed by families caring for mentally ill relatives are often significant, but are generally underestimated in cost models of studies of community care programs (Dickey, McGuire, Cannon, & Gudeman, 1986; Goldman, 1982).

There is some controversy as to whether to include research costs related to the alternatives being evaluated. Cannon (1985; Dickey, McGuire, Cannon, & Gudeman, 1986) consciously removed them from consideration in their cost-effectiveness studies of hospital versus residential care for chronic patients, as did Weisbrod (1983) in his analysis of inpatient versus community care for a more heterogeneous group of patients. This was done on the basis of the perception that research was a joint product of the production process, but did not contribute to the product of patient care. Conversely, Siegert and Yates (1980) included research costs in their cost-effectiveness study of three different service delivery systems for teaching child management skills to parents of troubled children, on the basis of empirical findings which demonstrate that research activities can enhance treatment effectiveness (Yates, 1985; Kiresuk & Lund, 1978). Similar arguments can apply to the inclusion or exclusion of costs for providing practica to university students in mental health disciplines. Perhaps these issues can best be resolved through further empirical verification of the extent to which training and research activities contribute to service outcomes. Nevertheless, in determining the applicability of findings from cost-benefit and cost-effectiveness analysis of social experiments to full-scale policy implementation, it seems necessary to separate developmental research costs from operating costs (Levin, 1983b).

There is also some question about whether to include the cost of transfer payments in a cost accounting developed from a social perspective. As indicated above, the theory of cost-benefit stipulates that transfer payments are "neither costs nor benefits . . . to the economy as a whole . . . , only a part of distributing the aggregate product" (Mishan, 1972, p. 60). However, in their studies of inpatient versus community care, Weisbrod (1983) and Cannon (1985) include such transfer payments as Supplementary Security Income (SSI) and Aid for the Families of Dependent Children (AFDC) as a proxy for the costs of subsistence in the community outside of the formal mental health system. This is justified on the basis of the expense and difficulty of measuring basic food, clothing and shelter expenditures directly, and implies the assumption that the transfer payments will be applied to subsistence expenditures on these commodities due to their low level. However, Rubin (1983) has noted that some portions of transfer payments are unrelated to maintenance of the individual client because they are conditional upon his or her family status and implied financial responsibilities. Furthermore, including all transfer payments as costs does not provide satisfactory data to compare rigorously the costs of maintenance in and out of the hospital.

Measurement

The simplest and least rigorous manner of measuring direct costs is to utilize the dollar charge for the service as a measure of the value of the resources utilized (Rubin, 1983). This is based upon the assumption of a perfectly competitive market in which the equilibrium price reflects the point at which marginal cost and marginal revenue are equal. It also ignores the established practice of health organizations in charging above the true marginal cost of some services to subsidize others (Warner & Luce, 1982). Beyond this, the charge may vary according to the peculiar conditions imposed by third party payers (Cannon, 1985).

A more rigorous approach to the measurement of direct services cost is based upon budgetary or expenditure data from the organization delivering the service, supplemented by additions for the value of goods and services which are provided free of charge and the opportunity cost of any portion of the land and physical plant which is owned by the organization. Levin (1983a; 1983b) has noted that expenditure data are superior to budgetary data in that the latter reflect estimated rather than actual costs. However, in the case of the public and non-proprietary organizations which deliver most mental health services (Weisbrod & Schlesinger, 1981), expenditure data may also present serious problems which require careful attention.

First, distortions may exist when the cost of major capital rehabilitation is allocated completely to the year in which the expenditure occurs, rather than being prorated over the expected life of the improvement. Second, the costs of the service of interest may have to be disentangled from joint products which result from the same production process. Third, an appropriate proportion of general administration and overhead costs of an organization producing many services may have to be determined and allocated to the service being evaluated. Fourth, the structure of the organization's financial records may render it difficult to determine the resources utilized in producing a particular service with any degree of specificity.

These and similar difficulties often limit the reliability and validity of cost measurements derived from the expenditure data of mental health agencies (Fishman, 1981). For example, Zelman, Stone and Davenport (1982) have demonstrated that it was possible to develop 17 different estimates of the cost of community mental health center children's service programs utilizing various practices which have been used in various studies. The estimates varied tremendously. Similarly, in another health setting, Finkler (1981) reported difficulties in replicating an earlier cost-effectiveness study of regional heart surgery centers because of differences in recording and retrieving cost data. Thus, a cost-benefit or cost-effectiveness study is often limited by the quality of organizational expenditure data and should clearly specify the procedures utilized to transform that data into costs.

The measurement of indirect resources provided by the patient, one's family, and peers generally includes out-of-pocket expenditures for transportation, recreational activities, and sometimes loss of earnings due to employment absences. Ideally, it should also include estimates of the amounts of their time and psychological resources which are consumed because of factors attributable to the alternatives being studied (Cannon, 1985; Levin, 1983a, 1983b; Weisbrod, 1983; Yates & Newman, 1980).

Valuation

In most cost-benefit and cost-effectiveness studies of mental health programs, it is assumed that with some important exceptions market prices appropriately reflect the value of the personnel, facilities, equipment, and supplies consumed in the direct provision of mental health services and the indirect provision of other social, health, and safety services related to the alternatives under study (Dickey, Cannon, & McGuire, 1983; Frank, 1981; Levin, 1983a). The first exception involves the possibility that the acquisition of a staff member possessing a skill which is in short supply may bid up salaries of similar personnel who are employed in other settings. Any rise in labor costs in the market in general which is attributable to the program under study should be considered as a cost of it.

The second exception involves the need to impute a shadow price for the value of volunteer labor. If acceptably proximate market analogues are available, the salaries and benefits generally paid to the analogues in the appropriate market can be utilized (Levin, 1983a; Warner & Luce, 1982). This would be a measure of what the costs to the program would be if the volunteers were replaced by paid staff. Yates (1985) has suggested the alternate procedure of valuing volunteers' time according to their actual employment earnings in the labor market. While this represents the opportunity cost of the time of particular volunteers, it seems somewhat arbitrary in that these earnings may not be related to the volunteers' activities in the program being evaluated.

The third exception relates to the need to apply the appropriate market price to donated equipment and supplies. In the case of equipment, this may involve the utilization of appropriate lease or rental costs observed in the local market. An alternate procedure is to estimate the replacement cost of the equipment and adjust it for depreciation and the opportunity cost of the imputed investment (Levin, 1983a).

Similarly, a fourth exception involves land and physical plants which are owned or utilized without charge by the program under study. This is especially important in the economic evaluation of mental health institutional programs utilizing publicly owned land and buildings, since these costs often constitute a significant proportion of the total resources utilized by these programs. The opportunity cost of the owned or donated land and facilities could be valued through determining the actual market rental or lease cost for similar land and physical plants (Levin, 1983a). However, this may be difficult in the case of facilities such as state mental hospitals for which it would be difficult to locate analogues which had been traded on the rental market.

The alternate procedure would involve determination of a realistic stock value of the capital and adjusting it for depreciation and the opportunity cost of the resources which had been invested in purchasing the facility. The accounting entries for stock values and depreciation costs of these facilities are often artificially low, as they are based upon historical construction costs which are far below replacement cost (Weisbrod, 1983). Cannon, McGuire, and Dickey (1985) and Cannon (1985) have developed a three-step procedure for imputing capital costs of publicly owned mental health facilities. The first step is to estimate the replacement cost of the facility as an upper bound of its stock value. The second step involves the development of a range of market values through consultation with real estate experts and observation of the purchase price of any similar facilities. The third step involves the conversion of the stock values into flow values by applying various rental rates reflecting differing assumptions regarding depreciation and rates of return on capital to them.

Indirect resources related to the time of families and other collaterals expended in performing services which are associated with the treatment alternatives under study can be valued according to the market wage rates which would have to be paid to non-volunteers to perform these functions (Warner & Luce, 1982). However, only rudimentary methods for measurement of the psychological costs to families and collaterals have been developed, and no acceptable techniques exist for their valuation (Dickey, McGuire, Cannon, & Gudeman, 1986; Thompson & Doll, 1982).

The value of indirect resources related to the time which clients expend in treatment and in waiting for treatment can be estimated through utilization of their after tax wage rate or the market wage rate of domestic employees for homemakers (Gramlich, 1981; Yates, 1985). This is somewhat problematic in that the time of unemployed clients who are not homemakers would be assessed as valueless. However, Weisbrod (1983) argues that this is a plausible valuation for the leisure time of unemployed severely disabled chronic clients, "given their emotional problems, the negative psychological feelings associated with leisure—not being with other people in a work setting, not doing work that receives social approval'' (p. 823). In many cases, the use of the after tax wage rate seems justifiable on pragmatic grounds, given the unavailability of other behavioral information about how clients or others value their time.

Ideally, cost-effectiveness and cost-benefit approaches to program evaluation should determine marginal as well as average costs in order to account for variations in returns to scale and the degree to which alternatives are operating at optimal output levels (Dickey, Cannon & McGuire, 1986). This is especially appropriate when the evaluation question involves the costs of serving additional individuals rather than the comparison of entire programs (Rubin, 1982). The description of the proportions of fixed and variable costs involved in each alternative is also useful in dealing with issues of scale and scope (Dickey, Cannon, & McGuire, 1986).

Finally, it has been demonstrated that various client characteristics are related to resource utilization (Bloom, 1970; Lavik, 1983; Henisz, Flynn, & Levine, 1981); and, therefore, the description of the cost of resources consumed by clients of various types is very useful in determining the effects of case mix on average cost differences among alternatives (McCaffree, 1968). This is extremely important since an alternative may be cost-beneficial or cost-effective for some types of patients and not others. However, cost data which is disaggregated according to client characteristics have often been unavailable even in the most sophisticated attempts to evaluate mental health programs economically (Weisbrod, 1983).

ANALYSIS OF BENEFITS

Benefits, or effects, are the outcomes of the production process, and their analysis encompasses identical processes to those utilized in the analysis of costs (Warner, 1983). In pure cost-benefit approaches all measured outcomes are valued in monetary terms, while cost-effectiveness analysis allows for the assessment of nonmonetized outcomes. As has been suggested above, in economic evaluations of mental health programs some outcomes such as after-tax wages of clients attributable to the program can be monetized, while others, such as clinical symptomatology, can be more meaningfully measured in non-monetary units. Rubin (1983) has noted that while methods for valuing human lives saved may be controversial, there is currently no available methodology at all for valuing days of improved mental health. Other indirect outcomes such as effects upon neighbors of clients are generally regarded as not measurable in specific quantitative terms, given resource and conceptual constraints (Weisbrod & Schlesinger, 1981; Panzetta, 1973). Even the measurement of property devaluation resulting from community-based services has proven difficult (Dear & Taylor, 1982). Therefore, a pure cost-benefit approach will often have the disadvantage of ignoring important program effects (Glass & Goldberg, 1977).

In classical cost-benefit approaches, benefits are determined through observation of competitive markets to ascertain whether projects provide increases in consumer surplus (Mishan, 1972). Consumer surplus refers to "the difference between the total value consumers receive from the consumption of a particular good or service and the total amount which they pay for it" (Nicholson, 1983, p. 689). In these terms, the benefits of a project are valued as the total amount which "individuals are willing to pay rather than go without it" (Stokey & Zeckhauser, 1978, p. 150).

This methodology is not appropriate for valuing the benefits of alternatives assessed in many cost-benefit analyses of mental health programs from a social perspective for the same reasons that markets cannot be relied upon to ensure efficient allocation. First, most mental health services are not allocated through a market mechanism, but are provided without charge or far below actual cost under public and non-proprietary auspices, sometimes to clients who are legally coerced to participate (Weisbrod & Schlesinger, 1981). Second, where mental health services are provided in a market context, the efficiency of allocation is distorted through the moral hazard which relates to the participation of third party insurers (McGuire, 1981). This moral hazard refers to incentives which insurance provides to policyholders to utilize quantities of service beyond those which would be consumed in responding to price in a competitive market. Thus, the consumers' observed willingness to pay may be an inflated indication of the actual value of the benefits to them. Third, ignorance, unavailability of information about the effectiveness of service, and, in some cases, the inability to make rational decisions may lead consumers to over or undervalue services in deciding upon the quantity to purchase at a given market price. Furthermore, these factors may subject them to a great deal of influence from service providers who may have incentives to induce demand (Arrow, 1963; McGuire, 1981; Weisbrod & Schlesinger, 1981). Finally, market prices may not reflect the external benefits enjoyed by family members, work colleagues, other community members, and society at large (Mishan, 1972; Romans, 1973), and thus a valuation of benefits relying strictly upon the clients' willingness to pay would tend toward underestimation. Therefore, economic evaluations of mental health programs must often go beyond the model of a perfectly competitive private market in articulating outcomes. Fortunately, some benefits such as enhanced employment are valued in markets which often do reflect their social value.

Romans (1973) has suggested that the cost of producing the same outcome through the best alternate means offers a minimum estimate of the benefits of a mental health program. However, this cost-effectiveness approach implies that some resources must be applied to ameliorating a particular mental health problem, even if their allocation to another goal is more efficient. This is most plausible in situations in which doing nothing about a problem will result in discernible risk to public safety.

DISCOUNTING OF COSTS AND BENEFITS

In many mental health programs, resources are consumed and benefits are created over multiple years. Furthermore, alternatives being compared may exhibit different time patterns in this regard. However, costs and benefits cannot simply be summed across years because, even exclusive of inflation, future dollars are less valuable than current dollars (Warner & Luce, 1982; Gramlich, 1981). This relates to the increased opportunities which currently accessible resources provide for consumption and investment. Thus, in a cost-benefit approach to program evaluation, the streams of costs borne and benefits realized over multiple years must be reduced to a common monetary base which involves determining their present values.

Determination of the present value of a program requires the identification of the value of resources consumed and benefits which occur in each year, and application of a discount rate to those which occur beyond the first year. In cost-benefit analyses of mental health programs, future benefits are sometimes estimated through extrapolation from cross sectional or experimental data from a rather brief period (Murphy & Datel, 1976). However, there is very limited empirical support for determining the longitudinal pattern of service utilization or the horizon over which benefits will continue to be realized, especially if the mental health intervention is varied (Stein & Test, 1980).

The selection from among alternate discount rates is often controversial and sometimes can profoundly effect the results of the analysis (Stokey & Zeckhauser, 1978). The discount rate for public sector activities should be reflective of the rate of return on society's best alternative use of its resources, which is the social opportunity cost of the program under study (Warner & Luce, 1982). Grossly, three alternate approaches are available (Gramlich, 1981; Mishan, 1972). The first involves the use of the opportunity cost for private sector capital, generally the gross before-tax rate of interest on private investment, sometimes adjusted for the social cost of externalities such as pollution. Its use is based upon the notion that public investment exactly displaces private investment and is most applicable when this has clearly occurred. The second approach involves the use of a weighted average of the gross before-tax rate of interest on private investment and the after-tax rate of return on private savings, and is based upon the notion that funds for public projects displace both private investment and private consumption. However, the after-tax rate of return on private saving may not reflect the marginal rate which savers require to compensate them for their savings due to market imperfections flowing from government financial regulation and taxes. The third approach is based upon the notion that the social rate of time preference should be lower than individual rates of time preference in order to consider the consumption of future generations. Thus, the social opportunity cost will generally be lower than the private sector opportunity cost which does not fully consider this factor. One variant of this perspective argues that the discount rate should be set at the economy's growth rate which reflects a country's optimal combination of investment and savings (Marglin, 1963).

As Nicholson (1983) has noted:

> In looking at the real world a government analyst is faced with a bewildering assortment of interest rates, ranging from rates on very liquid short-term assets (say 0 percent to 20 percent) through the rate of return on capital in very profitable industries (which may be higher than 40 percent). (p. 624)

When this is combined with the finding that "when we get down to discussing real numbers most of the voluminous discount rate literature becomes suddenly vague" (Stokey & Zeckhauser, 1978, p. 22), the importance of reflecting the implications of a range of rates for the findings of the analysis becomes clear. Stokey and Zeckhauser suggest that the interest rate on long-term government bonds provides a reasonable lower bound since a project should not be undertaken if it cannot yield a return as high as the cost to government of borrowed funds.

Discounting can clearly be applied to multi-year costs in a cost-effectiveness study as well as to multi-year costs and benefits in a cost-benefit study (Panzetta, 1973). However, Watts, Jackson, and Logerfo (1979) have argued that discounting should also be applied to non-monetized effectiveness measurements in order to transform them into comparable homogeneous units. In the absence of such discounting, the underlying implication is that a unit of outcome realized today is no more valuable than one produced in the future. Since costs are discounted on the basis that present dollars are more valuable than future dollars, an inherent incentive to defer production of non-monetized effects forever is created (Warner & Luce, 1982). While the logic for discounting non-monetized effects over time seems sound, there is little conceptual basis for determination of the specific rate (Watts, Jackson, & Logerfo, 1979).

In a cost-benefit analysis, once the discount rate or rates are determined, the net discounted present value of the program can be calculated as follows:

$$V_j = \sum_{t=1}^{T} \frac{B_t - C_t}{(1 + r)^t}$$

where: V_j = Net Discounted Present Value;
B_t = Benefits from the proposed program in year t;
C_t = Costs of the proposed program in year t;
T = The horizon over which all costs are borne and all benefits are realized; and
r = Discount rate

Thus, if the net discounted present value is larger than zero, the program alternative represents an efficient resource allocation, and the program with the highest net benefits is evaluated as the most desirable. Inflation can be dealt with either by utilizing constant dollars and the real interest rate for discounting, or nominal inflated dollars and the nominal interest rate for discounting (Gramlich, 1981).

The internal rate of return is the discount rate which renders the present value of benefits exactly equal to the present value of costs, and can be calculated by solving the above equation for r when Vj is equal to zero (Mishan, 1972). It defines the discount rate at which consumption of resources by the program alternative is indifferent on efficiency grounds. Thus, if the appropriate opportunity cost of capital is greater than the internal rate of return, the allocation would be inefficient.

The net discounted present value criterion is preferable to the internal rate of return in many situations, especially when alternatives with different horizons are being compared. The internal rate of return and present value criteria yield equivalent results only when budget constraints are not relevant, alternatives are not mutually exclusive, and streams of net returns are first negative and then positive (Gramlich, 1981; Stokey & Zeckhauser, 1978; Mishan, 1972).

CONCLUSION

It is clear that cost-benefit and cost-effectiveness analysis hold great promise as means of making relative valuations among alternatives studied in human service program evaluations. However, they also present unique methodological challenges in the analysis and discounting of costs and benefits. These challenges can be met with sufficient rigor when human service program evaluators combine their knowledge of service settings with an adequate understanding of the theory which underpins the cost-benefit framework.

REFERENCES

Arrow, K. (1963). Uncertainty and the welfare economics of medical care. *American Economic Review*, *53*, 941-973.

Bloom, B.L. (1970). Prediction and monitoring of resource utilization in a community-oriented mental hospital. *American Journal of Public Health*, *60*, 2257-2268.

Cannon, N.L. (1985). *A cost effectiveness analysis comparing treatment alternatives for chronically mentally ill patients*. Unpublished doctoral dissertation, Brandeis University, 1985.

Cannon, N.L., McGuire, T.G., & Dickey, B. (1985). Capital costs in economic program evaluation: The case of mental health services. In J.S. Catterall (Ed.), *New directions for program evaluation*. San Francisco: Jossey-Bass.

Cook, T.D., & Shadish, W.R. (1986). Program evaluation: The wordly science. *Annual Review of Psychology*, *37*, 193-232.

Dear, M.J., & Taylor, S.M. (1982). *Not on our street: Community attitudes to mental health care*. London: Pion.

Dickey, B., Cannon, N.L., & McGuire, T.G. (1986). Mental health cost studies: Some observations on methodology. *Administration in Mental Health*, *13*, 189-201.

Dickey, B., McGuire, T.G., Cannon, N.L., & Gudeman, J.E. (1986). Mental health cost models: Refinements and applications. *Medical Care*, *24*, 857-867.

Finkler, S.A. (1981). Cost-effectiveness of regionalization: Further results for heart surgery. *Health Services Research*, *16*, 325-333.

Fishman, D.B. (1981). *A cost-effectiveness methodology for community mental health centers: Development and pilot test*. Washington, DC: U.S. Government Printing Office.

Frank, R. (1981). Cost-benefit analysis in mental health services: A review of the literature. *Administration in Mental Health*, *8*, 161-176.

Glass, N.J., & Goldberg, D. (1977). Cost-benefit analysis and the evaluation of psychiatric services. *Psychological Medicine*, *7*, 701-707.

Goldman, H.H. (1982). Mental illness and family burden: A public health perspective. *Hospital and Community Psychiatry*, *33*, 557-590.

Gramlich, E.M. (1981). *Benefit-cost analysis of government programs*. Englewood Cliffs, NJ: Prentice-Hall, Inc.

Henisz, J.E., Flynn, H.R., & Levine, M. (1977). Clients and patients of mental health services. *Archives of General Psychiatry*, *34*, 1345-1348.

Kiresuk, T.J., & Lund, S.H. (1978). Goal attainment scaling. In C.C. Atkisson, W.A. Hargreaves, & M.J. Horowitz (Eds.), *Evaluation of human service programs*. New York: Academic Press.

Lavik, N.J. (1983). Utilization of mental health services over a given period. *Acta Psychiatrica Scandinavia*, *67*, 404-413.

Levin, H.M. (1983a). *Cost-effectiveness: A primer*. Beverly Hills, CA: Sage Publications.

Levin, H.M. (1983b). Cost-effectiveness analysis in evaluation research. In E.L. Struening & M.B. Brewer (Eds.), *The university edition of the handbook of evaluation research*. Beverly Hills, CA: Sage Publications.

Margin, S. (1963). The social rate of discount and optimal rate of investment. *Quarterly Journal of Economics*, *77*, 95-111.

McCaffree, K. (1968). The economic barriers to the development of community mental health programs. *Medical Care*, *4*, 286-299.

McGuire, T.G. (1981). *Financing psychotherapy: Costs, effects, and public policy*. Cambridge, MA: Ballinger Publishing Company.

Mishan, E.H. (1972). *Economics for social decisions*. New York: Praeger Publications.

Murphy, J.G., & Datel, W.E. (1976). A cost benefit analysis of community versus institutional living. *Hospital and Community Psychiatry*, *27*, 163-170.

Nicholson, W. (1983). *Intermediate micro-economics and its application* (3rd ed.). Chicago: The Dryden Press.

Panzetta, A.F. (1973). Cost-benefit studies in psychiatry. *Comprehensive Psychiatry*, *14*, 451-455.

Romans, J.T. (1973). The economic evaluation of mental health programs. *International Journal of Mental Health*, *2*, 38-50.

Rossi, P.H., & Freeman, H.E. (1985). *Evaluation: A systematic approach* (3rd ed.). Beverly Hills, CA: Sage Publications.

Rubin, J. (1982). Cost measurement and cost data in mental health settings. *Hospital and Community Psychiatry*, *33*, 751-754.

Rubin, J. (1983). Benefit-cost analysis and the care of the chronic psychiatric patient in the community. In I. Barofsky & R.D. Budson (Eds.), *The chronic psychiatric patient in the community: Principles of treatment*. New York: S.P. Publications.

Scriven, M.S. (1980). *The logic of evaluation*. Inverness, CA: Edgepress.

Sharfstein, S.S., & Clark, W. (1978). Economics and the chronic mental patient. *Schizophrenia Bulletin*, *4*, 399-414.

Siegert, F.A., & Yates, B.T. (1980). Behavioral child management cost-effectiveness: A comparison of individual in-office, individual in-home, and group delivery systems for behavioral child management. *Evaluation and the Health Professions*, *3*, 123-152.

Stein, L.I., & Test, M.A. (1980). Alternatives to mental hospital treatment: A conceptual model, treatment program, and clinical evaluation. *Archives of General Psychiatry*, *37*, 392-397.

Stokey, E. & Zeckhauser, R. (1978). *A primer for policy analysis*. New York: W.W. Norton and Company.

Thompson, E.H., & Doll, W. (1982). The burden of families coping with the mentally ill: An invisible crisis. *Family Relations*, *31*, 379-388.

Warner, K.E. (1983). Systematic analyses: Cost-benefit and cost-effectiveness techniques. In B.S. Bloom & S.E. Berki (Eds.), *Cost-benefit, cost-effectiveness, and other decision-making techniques in health care resource allocation. Proceedings of a regional symposium May 19-21, 1983, Chicago, Illinois*. New York: Biomedical Information Corporation.

Warner, K.E., & Luce, B.R. (1982). *Cost-benefit and cost-effectiveness analysis in health care: Principles, practice, and potential*. Ann Arbor, MI: Health Administration Press.

Watts, C.A., Jackson, M., & Logerfo, J.P. (1979). Cost-effectiveness analysis: Some problems of implementation. *Medical Care*, *17*, 430-434.

Weisbrod, B.A. (1983). A guide to benefit-cost analysis as seen through a controlled experiment in treating the mentally ill. *Journal of Health Politics, Policy and Law*, *7*, 808-845.

Weisbrod, B.A., & Schlesinger, M. (1981). Benefit-cost analysis in the mental health area: Issues and directions for research. In T.G. McGuire & B.A. Weisbrod (Eds.), *Economics and mental health*. Washington, DC: U.S. Government Printing Office.

Yates, B.T. (1985). Cost-effectiveness analysis and cost-benefit analysis: An introduction. *Behavioral Assessment*, *7*, 207-234.

Yates, B.T., & Newman, F. (1980). Approaches to cost-effectiveness analysis and cost-benefit analysis of psychotherapy. In G.R. Vandenbos (Ed.), *Psychotherapy: Practice, research, policy*. Beverly Hills: Sage Publications.

Zelman, W.N., Stone, A.V., & Davenport, B. (1982). Factors contributing to artifactual differences in reported mental health costs. *Administration in Mental Health*, *9*, 40-52.

Efficiency in Human Services: The Case of Education

David Stern, PhD

There are always plenty of proposals for improving education, but never enough money to implement most of them. Recommendations for raising teachers' salaries, reducing class size, reorganizing curriculum, and promoting new educational technologies all compete for the limited budgets available. The problem is how to get the most benefit from any given budget. That is what cost-effectiveness analysis attempts to determine.

EFFICIENCY AND INCENTIVES IN PUBLIC SCHOOLS

Economic analysis of education addresses two main questions. First, does the monetary value of benefits produced by expenditures on education equal or exceed the cost of those expenditures? Second, are schools and other educational efforts producing as much learning as possible, given their current budgets? The first question demands a *benefit-cost* (or cost-benefit) analysis. Examples of benefit-cost analysis in education are given below. Most of the attention in this paper will focus on the second question, which underlies *cost-effectiveness* analysis.

If education were a commodity like breakfast cereal or a service

---

Dr. Stern is Associate Professor, School of Education, University of California at Berkeley, Berkeley, CA 94720.

This article contains sections written by David Stern from a paper by David Stern and Guy Cox, entitled, "Assessing Cost-Effectiveness of Computer-Based Technology in Public Elementary and Secondary Schools," prepared for the Office of Technology Assessment, U.S. Congress, 1987.

like hair styling, economists would not ask these questions. Benefit-cost analysis would not be necessary because the operation of markets tends to equate the cost of producing a box of cereal or a haircut with its value to the consumer, measured by his or her willingness to pay. In other words, consumers do their own benefit-cost analysis, and this leads to efficient allocation of resources, since members of society are not seriously affected by each other's choice of breakfast cereal or hair style.

However, education is obviously different from these private goods and services, since members of society are seriously affected by each other's education. Educating one person benefits other members of society because education improves that person's competence as a citizen, parent, and neighbor. Adam Smith himself urged government support of local schools for this reason. Education also enables some individuals to make cultural and scientific contributions, for which the material compensation they receive is less than the benefit to society as a whole. Alfred Marshall saw this as a reason for public support of schools. Most contemporary economists still agree that education produces non-marketable benefits, and therefore warrants some degree of public funding because people would buy too little of it otherwise (Benson, 1978; Cohn, 1979; Haveman & Wolfe, 1984). For these reasons, nations everywhere in the world today not only support education with public funds but even require school attendance by law.

Compulsory schooling laws are intended to give people less choice about education than they have about haircuts or breakfast food. In addition, using government-operated schools as the principal mechanism for channeling tax revenues into education further restricts choice. At the elementary and secondary levels in the United States, the only way for households to obtain the maximum subsidy for schooling is to send their children to public schools. Although private schools are exempt from paying taxes, for the most part their expenses are not paid by government. If some sort of voucher scheme made it possible for private schools to be supported fully by public revenues, more households probably would choose private schools. At present, only about ten percent do so.

Given that public schools face little competition at the elementary and secondary levels, economists are not surprised to learn that these schools generally do not appear to operate in a cost-effective manner. More than a hundred studies have produced empirical evidence bearing on the cost-effectiveness of public schools. The nature of these studies will be described below. Hanushek (1986), surveying this evidence, concluded:

> If we think of schools as maximizing student achievement, the preceding evidence indicates that schools are *economically inefficient*, because they pay for attributes that are not systematically related to achievement. (p. 1166)

This "does not come as a great surprise" to Hanushek or other economists (Levin, 1976) because public schools are not like business firms that lose revenues and eventually go out of business if they fail in market competition. Public school revenues come from local, state, and federal sources. These payments are not conditioned on any measure of what students learn. Local elections concerned with school taxes may sometimes be influenced by perceptions of school quality, but this is a highly imperfect mechanism. Furthermore, public expenditures for education tend to be driven by perceived needs, so that states and the federal government tend to add money when failures become more evident. In addition to programs serving special needs, states also distribute a large amount of money simply on the basis of student headcount. This creates some incentive to maintain attendance, but not to increase learning.

This line of thinking has led some economists and others to argue for education vouchers, so that a school would receive less revenue if fewer families chose to attend. A less sweeping kind of proposal, also intended to make public schools more accountable, is to declare a school educationally bankrupt when it grossly fails to meet its obligations to students (Committee for Economic Development, 1985). A bankrupt public school would then be taken over and run by the state. Recently, the National Governors Association endorsed a proposal like this. These proposals are prompted by the evident lack of incentive for public schools to operate cost-effectively.

Also evident is the lack of material incentive for individual teachers and administrators to produce as much student learning as possible. Students have to be active collaborators in the process, of course, and motivating students who are required to be there is part of the challenge for teachers and administrators. How well they surmount this challenge and find ways for students to learn does not determine how much the teachers or administrators are paid. This evident lack of material incentive to do the job as well as possible has inspired many experiments with "merit pay" for teachers (Hatry & Greiner, 1985; Cohen & Murnane, 1985). In recent years, career ladders have been tried out in various places to give teachers more extrinsic incentive.

What has destroyed or distorted many merit pay plans in the past is the difficulty of defining and measuring what makes some teachers more effective than others. Simply relying on supervisors' ratings leads to charges of unfair subjectivity, but developing other kinds of evidence (dossiers, observations, and review by committees) tends to become inordinately time-consuming. Recently, some districts have tried awarding bonus pay for teachers in groups rather than as individuals. Group incentives are intended to avoid competition between individuals, and foster collaboration instead. Another idea receiving much current attention is to create a national board that would certify teachers for both basic and advanced levels of competence (Carnegie Forum, 1986). The prototype examinations now being developed for this proposed national board presumably will embody some conception of what good teachers do.

For the time being, teachers and administrators lack clear external standards of practice, either as individual members of a profession or as employees in a bureaucratic organization. Teachers and administrators who work hard do so because they are intrinsically motivated, and for this parents of public school children can only be grateful.

Whether it is possible to create extrinsic incentives for school professionals that reinforce their intrinsic motivation, and do not lead merely to teaching the test or other counterproductive games, remains an open question (Stern, 1986). But the fact remains that decision-making in public schools has not been systematically oriented toward maximizing students' learning.

One implication of not explicitly seeking to maximize students' learning is the lack of systematic innovation by schools themselves.

Schools are constantly innovating, but they generally do not keep track of the results. Innovations therefore tend not to produce cumulative understanding as fast as might happen if innovation were coordinated and monitored more systematically. In particular, schools almost never examine the relative cost-effectiveness of different techniques for achieving given objectives. Empirical studies of cost-effectiveness in education have been done mainly by researchers outside the schools themselves.

Another implication of schools' lack of systematic incentive to maximize students' learning is that results of cost-effectiveness studies by outside researchers or agencies will not necessarily influence what schools do. Change itself is costly and time-consuming. What seems to make sense to an outside analyst does not necessarily persuade insiders to act. As we analyze cost-effectiveness, therefore, we should try to see it from the point of view of each participant in the educational process.

BASIC PARADIGM FOR ANALYZING COST-EFFECTIVENESS

Economists analyzing cost-effectiveness in education begin with a basic theory of the "firm." A firm is a single decision-making entity which uses one or more inputs, or "factors of production," to produce one or more outputs. Inputs include such things as the time of various kinds of employees, for which the firm must make explicit payments. Inputs also include equipment or facilities which the firm may own but for which it incurs an "opportunity cost," since physical capital could be liquidated and the money invested in other ways. Outputs include goods or services for sale.

Profit is the difference between the revenues obtained from sale of outputs and the cost of producing those outputs. If a firm is maximizing profit, input quantities must be adjusted such that the value of output produced by the last unit of each input is equal to the unit price of that input. (For a standard derivation of this result, see Nicholson, 1983.) For instance if a dentist's office could increase its revenues by \$30 as a result of hiring a hygienist for one additional hour, and if the cost to the office for that additional hour were only \$15, then the office could increase profit by increasing hygienist's time. If the office added more and more hygienist's time, eventually space limitations and other constraints would reduce the amount of additional revenues from an additional hour of hygienist's time to $15, which equals the hygienist's hourly wage. At that point, if nothing else changed, the office would be maximizing its profit from hygienists.

A firm is said to be economically inefficient if it does not maximize profit, given available technology. There are two distinct ways in which a firm may be inefficient. First, it may fail to obtain the maximum feasible output from the inputs it is using. This is called technical inefficiency (Levin, 1976; Jamison & Lau, 1982). Simply wasting time or materials, due to ignorance or carelessness, creates technical inefficiency.

A firm may be technically efficient but still fail to maximize profit if it buys the wrong combination of inputs or sells the wrong amounts of output. This is called allocative inefficiency. Technical inefficiency can be observed in the physical process of production itself, without knowing the prices of inputs and outputs. In contrast, measuring allocative inefficiency requires information about what various inputs cost and what price can be obtained for output. In other words, technical efficiency is the absence of physical waste; allocative efficiency requires a correct response to prices. Both are necessary for profit maximization.

This basic paradigm for analyzing efficiency can be applied to public schools, even though we cannot compute a school's "profit" when students do not pay tuition and there is no explicit price for students' learning. It is possible to compare schools on the basis of either technical or allocative efficiency.

Educational Production Functions

Most of the research by economists has focused not on schools' technical efficiency, but on their allocative efficiency. To test whether schools are efficient in allocating their budgets, economists have compared the cost of various inputs with their estimated productivity. Productivity is usually estimated from a multiple regression in which measures of students' background and prior achievement, along with school inputs, predict students' current achievement. These regression models are called "educational production functions." The research on educational production functions has been reviewed by Lau (1979), Benson (1978), Cohn (1979), and Hanushek (1986), among others.

Coefficients on school inputs in an estimated production function can be compared directly with the cost to schools of purchasing these inputs. For example, Levin (1970) used a production function (estimated by Hanushek, 1968) in which teachers' score on a test of verbal ability and teachers' years of experience were included, along with other school inputs and characteristics of students, as predictors of achievement in sixth grade. Levin also estimated the additional salary cost associated with higher verbal scores and years of teaching experience. (Standard salary schedules for teachers award higher pay automatically as experience increases through the first two decades or so. Verbal or other test scores are not explicit factors on salary schedules, but the higher cost of recruiting teachers with higher test scores was estimated by Levin and has also been substantiated by other studies, e.g., Manski, 1985.) Comparing the cost of these two characteristics of teachers with their estimated effect on students' achievement, Levin discovered that the annual cost of producing a one-point gain in a sixth-grade student's achievement was $26 (in 1968 dollars) when the gain was obtained by employing teachers with higher verbal scores, while producing the same gain by employing teachers with more experience would cost five to ten times as much. This indicates that schools could produce higher student achievement—with a given budget—by recruiting teachers with high verbal scores instead of more experience.

Hanushek's (1986) recent review of more than a hundred production function studies came to the general conclusion, quoted above, that public schools are generally inefficient because they pay for inputs which do not seem to produce achievement by students. Specifically, effects of the following inputs on students' achievement have *not* usually been found to be positive and statistically significant: higher educational attainment by teachers (which leads to higher pay on the standard salary schedule), longer teaching experience, smaller class size, higher teachers' salary, or larger expenditures per pupil. The implication is that spending less on these inputs would do no harm and would therefore be cost-effective. Naturally, educational practitioners emphatically reject this conclusion.

There are several well-known reasons why "educational production functions" should not be taken at face value. These reasons have all been carefully stated by contributors to, or critics of, this research in its early stages, during the aftermath of the 1966 Coleman report (U.S. Office of Education, 1970), and repeated by reviewers already cited. Well-known limitations include the difficulty of imputing causality in correlational studies (e.g., do more experienced teachers somehow teach better, or are they simply in a better position to choose the students they teach?); inadequacies of certain available data (e.g., the survey analyzed in Coleman, 1966, did not measure change or match individual students to individual teachers); and difficulty of analyzing more than one output at a time.

In addition, the informal meta-analytic procedure employed by Hanushek (1986) yields somewhat ambiguous results. For instance, his table (p. 1161) shows 60 studies had findings about teachers' salary and student achievement, of which only ten are said to be statistically significant. Of these ten, nine studies showed a positive association. Among the other 50, 15 were positive, 11 negative, and 24 had unknown sign. If the "true" association between salary and student achievement were zero, how likely would this pattern of results be? Applying more formal procedures (Hedges and Olkin, 1985) might not support Hanushek's conclusion that teacher salary makes no difference.

On the effect of class size, another variable which makes no difference according to Hanushek, formal meta-analysis by Glass (1984) did reach the opposite conclusion. Using more than a hundred comparisons produced by fourteen experiments in which students were randomly assigned to classes of different sizes, Glass fitted a logarithmic relationship between class size and achievement. He found that cutting class size in half (from 20 to 10, 30 to 15, 10 to 5, or whatever) on average increases achievement by 0.2 standard deviations, or approximately two grade equivalent months.

The purpose of estimating educational production functions is to predict the change in output consequent upon a change in one input at a time. Multiple regression is the statistical procedure most commonly used. If Hanushek or anyone else wanted to do a formal meta-analysis of this research, he or she would have to solve a technical problem: the fact that the coefficient on one predictor (e.g., teacher salary or class size) depends on what other predictors are in the equation. Since different studies use different combinations of predictors, averaging estimated coefficients across studies should take this into account. Some progress in developing meta-analytic procedures for multiple regressions is reported by Ash (1985).

Experimental Studies

Estimation of educational production functions has relied on naturally occurring variation in inputs and outputs among schools. As mentioned, causal interpretation of such correlational studies is often problematic. A more direct approach to estimating cost-effectiveness of school programs, which yields clearer causal inferences, is to conduct actual experiments. Educational researchers have reported on thousands of experimental or quasi-experimental studies, but only a few of these contain information on costs.

One set of experimental studies that kept careful account of costs was reported by Jamison, Fletcher, Suppes, and Atkinson (1976). Their report dealt with computer-assisted instruction (CAI) as a method of compensatory education for disadvantaged children. Achievement of students receiving CAI was compared with similar students who received no CAI. (The report implies that, in two of the experiments, comparison groups did receive compensatory education in some other form.) CAI was found to produce statistically significant gains. Moreover, the per-pupil cost of CAI was found to be well within the per-pupil budget available for compensatory education. The authors concluded, therefore, that CAI was feasible and (if comparison groups were receiving conventional compensatory education services) cost-effective as a form of compensatory education. They also computed how big an increase in class size would be required if compensatory funds were not available and CAI had to be financed by reducing the number of teachers. They conclude the CAI would be cost-effective as a partial replacement of teachers. However, this part of their analysis assumed class size does not affect achievement – a debatable assumption, given the analysis by Glass (1984) described above.

More recently, Levin, Glass, and Meister (1984) (also, Levin & Meister, 1986; Levin, Glass, & Meister, 1986) have combined correlational, experimental, and quasi-experimental findings in order to compare the cost-effectiveness of four different educational interventions: reducing class size, lengthening the school day, introducing CAI, and instituting a cross-age tutoring program. To estimate effects of reducing class size, the authors relied on Glass' (1984) meta-analysis of experimental studies, described above. Effects of a longer school day were estimated from results obtained by the Beginning Teacher Evaluation Study (Denham & Lieberman, 1980), in which teachers' and students' use of time was found to be correlated with students' gains in achievement. The estimated effect of CAI came from a carefully conducted experiment in Los Angeles, using drill-and-practice software produced by the Computer Curriculum Corporation (Ragosta, Holland, & Jamison, 1982). To estimate effects of a cross-age tutoring program, Levin et al. used quasi-experimental evaluation results from an exemplary program in Boise. They summarized their results in Table 1.

This shows CAI is cost-effective at the elementary level compared to increasing instructional time, but reducing class size appears to be more cost-effective than CAI in mathematics, and the student-tutor component of the cross-age tutoring program seems to be the most cost-effective of the four innovations, in both mathematics and reading.

Niemiec, Blackwell, and Walberg (1986) criticized the study by Levin et al. for estimating the effects of CAI and tutoring based on single cases, instead of synthesizing estimates from the dozens of cases which have been published. In response, Levin et al. (1986) argued that their purpose was to compare effect size with the cost of a well-defined program. Previously published studies of CAI and student tutoring have not included information on cost. Therefore, they maintained it was more appropriate to use data on estimated effects only from the single cases for which they had obtained cost data.

The kind of analysis Levin et al. have done yields direct and obvious implications for policy and practice. Replicating this analy-

Table 1

Estimated Effectiveness of Four Educational Interventions in Months of Additional Achievement Gain per Year of Instruction for Each $100 per Student

	Mathematics	Reading
CAI	1.0	1.9
Cross-age tutoring		
Peer component	4.6	2.2
Adult component	0.8	0.5
Increasing instructional time	0.5	1.2
Reducing class size		
From To		
35 30	1.4	0.7
30 25	1.2	0.6
25 20	1.0	0.5
35 20	1.1	0.6

(from Levin & Meister, 1986, p. 748)

sis with data from other cases would be highly desirable. The method could also be extended to account more explicitly for use of students' time, and to deal with multiple outcomes.

Benefit-Cost Analysis

As explained above, cost-effectiveness studies are concerned with whether schools produce the maximum amount of student learning, given limited budgets. In contrast, benefit-cost analysis asks whether the monetary value of benefits from education exceeds the amount of money spent on it. This is a question of external efficiency. Should expenditure on education increase or decrease, as opposed to the question of internal efficiency addressed by cost-effectiveness studies, and how should the given budget be spent?

The monetary value of benefits from education has been measured mainly by increased earnings of individuals who have completed more years of schooling. Education is seen as an investment in "human capital" (Becker, 1975). The cost of producing human capital includes teachers' salaries and other direct costs of schooling. The cost also includes students' time, which has an opportunity cost especially for older students, because they could be earning money if they were not in school.

By comparing the cost of the investment with the additional earnings obtained as a result of additional schooling, it is possible to compute the rate of return on the investment. This has been done for various levels of schooling, at different times and in various places (Psacharopoulos, 1985). The social rate of return counts public expenditures on schooling as part of the cost, and includes additional tax payments by more highly educated individuals as part of the benefit to society. In less developed countries, real social rates of return exceed 25% at the primary level; at the secondary and post-secondary levels the rates of return are around 15%. In the industrialized countries, real social rates of return average around 10% at the secondary and post-secondary levels.

Benefit-cost analysis can also be used to evaluate particular educational programs. For example, Weisbrod (1965) evaluated a high school dropout-prevention program; Hu, Lee, and Stramsdorfer (1971) analyzed the additional costs and benefits of high school vocational education; Garms (1971) evaluated a program (Upward Bound) for encouraging low-income high school students to attend college; Mallar (1982) and others have analyzed Job Corps training; Psacharopoulos and Loxley (1985) and Psacharopoulos (1986) have compared vocational versus general academic curricula at the secondary and post-secondary levels. All these studies have used earnings as the measure of benefit from additional schooling.

The use of earnings to measure benefit from schooling is based on human capital theory, which assumes that schooling enables individuals to earn more money because it actually renders them more productive. Critics have questioned this assumption, arguing that schools may merely sort and screen individuals, awarding degrees and diplomas only to those who were most able at the outset. Arrow (1973) and Spence (1973) showed it is conceivable that more able individuals find it profitable to attend school even if school does nothing to make them more productive; in this extreme case, the true social return to schooling would be negative. However, more recent theoretical models suggest that certain kinds of screening or credentialing mechanisms may actually induce too little investment in schooling rather than too much (Weiss, 1985). Empirical attempts to measure the screening effect have indicated that it accounts for some, but probably less than half, of the private return to schooling (Cohn, 1979). It is difficult to test directly whether schooling increases individuals' productivity, because in most industrialized work settings the interdependence of jobs makes it impossible to measure one individual's output separately. The clearest evidence that education does increase productivity has come from studies of individual farmers in less developed countries (Jamison & Lau, 1982).

To the extent that schools merely screen rather than teach, the additional earnings associated with further schooling overstate the social benefit from education. On the other hand, education also produces a number of benefits other than earnings. These are not easily translated into monetary values, and are therefore not included in quantitative benefit-cost calculations. Nonetheless, they are real. Some of these non-monetary benefits from education accrue mainly to the people who get the schooling. Better health is an example: there is evidence that more highly educated individuals are better able to keep themselves healthy (Haveman & Wolfe, 1984). Another example is the pure "consumption" benefit of knowing more about the arts and sciences. In addition to such benefits accruing to educated individuals themselves, other members of society also benefit. As mentioned above, we rely on each other's competence not only at work but also outside the sphere of market activity. And education enables some people to make contributions to the culture (e.g., basic research) whose value exceeds what the contributors are paid.

MULTIPLE OUTCOMES AND LIMITED TIME

The existence of multiple outcomes complicates the analysis of efficiency in education, as in other human services. Given that students and teachers have limited time, spending more time on one kind of learning must mean less time for other things. Therefore, any innovation that changes the use of time will generally have multiple effects. Students' performance may improve in the subject targeted by the innovation, but performance in other domains may decline. One apparent case of such effects was the experiment in educational performance contracting conducted in 1970-71. Students taught by contractors achieved slightly bigger gains in the subjects on which payments to contractors depended, but students in control groups did better in non-contracted subjects (Gramlich & Koshel, 1975).

Even when incentives to emphasize certain outcomes are not as explicit as in the performance contracting experiment, they do exist. Innovations are usually designed to achieve certain objectives more than others. For example, it is possible to design a CAI package to teach a certain narrow set of skills, then test improvement in these skills to measure the effect of CAI. If the CAI program took a substantial amount of students' time but the evaluation did not also consider other outcomes, it would be quite misleading.

Glass discussed the issue of time in his analysis of the CAI study in Los Angeles conducted by the Educational Testing Service (Ragosta, Holland, & Jamison, 1982, Part 6). The treatment in this study consisted of daily drilling-and-practice sessions conducted in a lab outside the regular classroom. Sessions lasted 10 or 20 minutes. Within a given classroom, students were randomly assigned to receive CAI in math only, reading and language only, or a combination of math with reading and language. While students assigned to math CAI were at the lab doing math, other students from the same classroom would either be in the lab doing language or reading, or they would be back in the regular classroom doing whatever the teacher's plan called for—sometimes math, but more often other subjects, since math does not take up most of the regular day in elementary school. As a result, students assigned to math CAI would spend more time on math than other students. Glass estimated that students receiving 20-minute daily sessions in math CAI spent an extra 50 hours a year on math, over and above the approximately 90 hours a year of math instruction received by other students. That is a large increment of time.

Extra time in one subject implies less time for other subjects. If performance in these other subjects declined, the innovation would have to be considered less successful than if such a decline did not occur (or was ignored). Unfortunately, most evaluations of CAI have not reported any evidence, one way or the other, on performance in subjects not specifically taught by the CAI itself. The ETS-LA study is an exception, since all students were tested in a range of subjects. However, in the ETS-LA study, students who received CAI only in math served as the controls for students who received CAI only in language and reading. Therefore, the design of the ETS-LA study makes it difficult or impossible to distinguish between (1) students doing worse in subjects where they received no CAI (and therefore had less time allocated), as opposed to (2) students doing better in subjects on which they received CAI (and spent extra time).

Some innovations are more subject-specific than others. Tutoring, like CAI, usually focuses on one or two subjects. In contrast, the effect of reducing class size, extending the school day, or raising teachers' salaries could be spread over all subjects.

Recognition of time constraints implies that an innovation should be evaluated on a broad range of outcomes if it changes students' use of time. If CAI or tutoring increases the amount of time students spend on a particular subject, and therefore reduces time on other subjects, then it may be misleading to use performance in that subject alone as the criterion for comparing the CAI or tutoring program against "traditional" instruction, or against some other innovation that has a diffuse but potentially positive impact on performance in all subjects. Taken to an absurd extreme, it would be possible to have students spend all their time practicing a particular skill such as long division, then give a test on long division only, and find a large "effect size" compared to students who had not spent all their time on that narrow subject. To avoid such misleading conclusions, evaluations have to consider the whole range of subjects students are supposed to be learning.

Cost-Effectiveness and Multiple Outcomes

When multiple outcomes are measured, analyzing cost-effectiveness becomes more complicated. However, new techniques are beginning to make this possible.

To illustrate the problem, suppose two programs are to be compared: CAI and a cross-age tutoring program. Suppose that there is a control group for each program, and a broad spectrum of outcomes are measured for students in each program and each control group. Suppose further that CAI is found to produce bigger gains than the tutoring program in math and in students' motivation to do schoolwork, but the tutoring program produces bigger gains than CAI in reading and in higher-order thinking. To make the comparison as simple as possible, suppose the two programs have equal cost. Which is more cost-effective?

One approach to this problem is to combine different outcomes into a single score. If the combination procedure is linear, this means assigning fixed weights to the different outcomes, giving more weight to outcomes that are considered more important. The resulting score then can be treated as a single outcome, and cost-effectiveness comparisons can proceed as usual. This is what Levin (1983) calls "cost-utility analysis."

Assigning weights to different outcomes requires, first, a procedure to select judges and, second, a procedure to elicit their judgments about the relative values of different outcomes. The first procedure is political. If the weights attached to different outcomes are to represent preferences of some constituency, representatives of that constituency have to be included. The constituencies interested in public schools include students, parents, teachers, administrators, taxpayers, and employers. A legitimate set of weights on different outcomes would have to represent these different groups. The judges who assign these weights would have to be selected accordingly.

Once a sample of judges is chosen, there are various possible procedures for determining weights. A simple procedure would have each judge assign a number—from one to ten, one to a hundred, or whatever—to each outcome, where a larger number signifies greater importance. These numbers can then be simply averaged across judges to derive weights.

An example of a more complicated procedure for deriving weights is "conjoint analysis" (Rao, 1977). This has been used in marketing research to discover what differences between competing brands are important to consumers. A sample of respondents is asked to make hypothetical choices – in this case, the choice would be among a set of hypothetical programs. Each program would be described in terms of outcomes – high on some, low on others. Respondents' ratings of programs can be analyzed to provide numerical weights signifying each outcome's importance. Results would probably be more accurate, in the sense of predicting how people choose schools, than would results from the simpler weighting procedure. However, conjoint analysis is usually done with only three or four dimensions to be considered. In education, it would be difficult to reduce the number of dimensions to less than a dozen.

Assigning fixed, numerical weights to different outcomes is therefore not a satisfactory solution. Simple weighting schemes are imprecise and somewhat arbitrary. More sophisticated methods may not be feasible and are also somewhat arbitrary.

There is an alternative solution which takes multiple outcomes into account but does not impose arbitrary, fixed weights on different outcomes. Charnes, Cooper, and Rhodes (1978) have developed a linear-programming approach to this problem, which they call "data envelopment analysis" (DEA). As they have described it, the purpose is to determine which "decision-making units," among a set of similar decision-making units, can be considered efficient. Schools are an example of decision-making units. Bessent and Bessent (1980) have, in fact, demonstrated the empirical application of DEA to schools in Houston, Texas.

DEA begins with an efficiency index, which can be computed for each decision-making unit. The index is the ratio of a weighted sum of outputs to a weighted sum of inputs. A different set of weights is computed for each decision-making unit, to maximize its efficiency index (constrained not to exceed unity). If the maximum efficiency index computed for a given decision-making unit, using the set of weights that make it look as efficient as possible, is less than the efficiency index computed for another decision-making unit using the same set of weights, then the first decision-making unit is deemed relatively inefficient.

If the purpose is to compare programs such as CAI or tutoring, instead of comparing decision-making units such as schools, then DEA could be modified as follows. Suppose we have data on dozens or hundreds of programs. Each program is some kind of intervention or alteration in a school or set of schools. From each program the data include a set of outcomes for students in the program compared to a control group. Suppose we also know the cost of each program. Then, using an algorithm like DEA, we could compute the weights on different outcomes that would yield the largest possible efficiency index (weighted effect size per dollar) for each program. A program would be considered inefficient if its efficiency index, using its "own" weights on different outcomes, is smaller than the efficiency index of some other program, computed with the same weights.

Figure 1 illustrates what could be learned from this procedure. A, B, C, D, and E are five programs. For simplicity, we consider only two outcomes, and assume here that all five programs cost the same amount of money per student. Among the five programs, A, B, and C would have maximum efficiency ratings, but D and E would not. The lines connecting efficient programs represent what economists

Figure 1. Production Possibilities Frontier

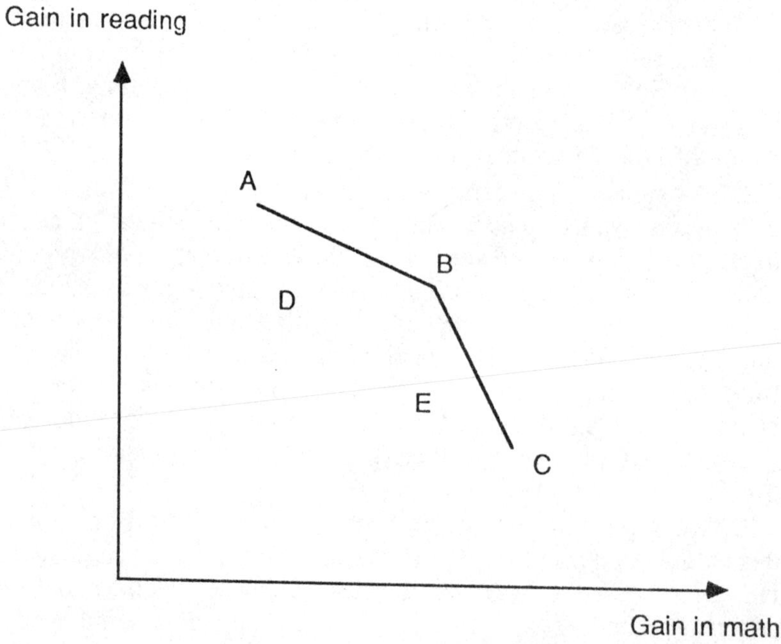

call the production possibilities frontier. Programs A, B, and C are on the frontier, but D and E are not.

Procedures like DEA have certain drawbacks. One is that results might change drastically if only one or a few observations were excluded from the analysis. For instance, in the diagram, program E would appear efficient if program C were not included. In addition, errors in measurement can produce large distortions in DEA results. Sexton, Silkman, and Hogan (1986) discuss these and other drawbacks of DEA, and suggest possible modifications of the technique. Standard economic procedures for estimating production functions may be more robust against errors in data or model specification, but these standard techniques do not handle more than one output at a time (Forsund, Lovell, & Schmidt, 1980; Levin, 1974). It is possible to measure relative efficiency of programs producing multiple outputs by using a cost function instead of a production function (Cavin & Stafford, 1985), if different programs pay different prices for a given input such as teachers' time.

In sum, it is possible to define and measure cost-effectiveness when multiple outcomes are considered. This would enable us to examine educational innovations or interventions at different times and places, and to decide which ones have been relatively efficient. The result might be that some CAI programs appear efficient and others do not, some tutoring programs are efficient while others are not, and similarly for other types of programs. Generalizations about the cost-effectiveness of one *type* of program (e.g., CAI) relative to another (e.g., tutoring), would be built upon analysis of many specific instances.

CONCLUSION

Cost-effectiveness analysis can point the way to doing more good with any given budget. In education, cost-effectiveness studies have yielded useful results. There has been some progress in measuring cost-effectiveness of programs that produce multiple outcomes. It would be feasible for school systems to monitor their own cost-effectiveness as a routine matter, but this will not happen unless schools are given the incentive to become more efficient.

REFERENCES

Alkin, M.C. (1969). *Evaluating the cost-effectiveness of instructional programs*. Berkeley: Center for the Study of Evaluation of Instruction, University of California.

Arrow, K.J. (1973). Higher education as a filter. *Journal of Public Economics*, *2*, 193-216.

Ash, M.S. (1985). *Meta-analysis and research integration of regression studies*. Unpublished doctoral dissertation, University of California, Berkeley.

Bangert-Drowns, R., Kulik, J.H., & Kulik, C.C. (1985). Effectiveness of computer-based education in secondary schools. *Journal of Computer Based Instruction*, *12*, 59-68.

Becker, G.S. (1975). *Human capital* (2nd ed.). Chicago: University of Chicago Press.

Benson, C.S. (1978). *The economics of public education* (3rd ed.). Boston: Houghton Mifflin.

Bessent, A.M., & Bessent, E.W. (1980). Determining the comparative efficiency of schools through data envelopment analysis. *Educational Administration Quarterly*, *16*, 57-75.

Capper, J., & Copple, C. (1986). Computers in education. *The Research into Practice Digest*, *1*(3), 1-25.

Carnegie Forum on Education and the Economy. (1986). *A nation prepared teachers for the 21st century*. Report of the task force on teaching as a profession. New York: Carnegie Corporation.

Cavin, E.S., & Stafford, F.P. (1985). Efficient provision of employment service outputs: A production frontier analysis. *Journal of Human Resources*, *20*, 484-503.

Charnes, A., Cooper, W.W., & Rhodes, E. (1978). Measuring the efficiency of decision-making units. *European Journal of Operational Research*, *2*, 429-444.

Cohen, D.K., & Murnane, R.J. (1985). The merits of merit pay. *The Public Interest*, No. 80, 3-30.

Cohn, E. (1979). *The economics of education* (rev. ed.). Cambridge, MA: Ballinger.

Coleman, J.S. et al. (1966). *Equality of educational opportunity*. Washington, DC: U.S. Government Printing Office.

Committee for Economic Development. (1985). *Investing in our children*. New York: Committee for Economic Development.

Denham, C., & Lieberman, A. (Eds.). (1980). *Time to learn*. Washington, DC: U.S. Department of Education.

diSessa, A.A. (1986). *The third revolution in computers and education*. Berkeley: Graduate School of Education, University of California.

Forsund, F.R. et al. (1980). A survey of frontier production functions and of their relationship to efficiency measurement. *Journal of Econometrics*, *13*, 5-25.

Garms, W.I. (1970). A benefit-cost analysis of the upward bound program. *Journal of Human Resources*, *6*(2), 206-220.

Glass, G.V., McGaw, B., & Smith, M.L. (1981). *Meta-analysis in social research*. Beverly Hills, CA: Sage Publications.

Gramlich, E.M. (1981). *Benefit-cost analysis of government programs*. Englewood Cliffs, NJ: Prentice-Hall, Inc.

Gramlich, E.M., & Koshel, P.P. (1975). *Educational performance contracting*. Washington, DC: The Brookings Institution.

Hanushek, E.A. (1968). *The education of negroes and whites*. Unpublished doctoral dissertation, Massachusetts Institute of Technology.

Hanushek, E.A. (1981). Throwing money at schools. *Journal of Policy Analysis and Management*, *1*, 19-42.

Hanushek, E.A. (1986). The economics of schooling. *Journal of Economic Literature*, *24*(3), 1141-1177.

Hatry, H.P., & Greiner, J.M. (1985). *Issues and case studies in teacher incentive plans*. Washington, DC: The Urban Institute.

Haveman, R.H., & Wolfe, B.L. (1984). Education, productivity, and well-being: On defining and measuring the economic characteristics of schooling. In E. Dean (Ed.), *Education and economic productivity*. Cambridge, MA: Ballinger.

Hedges, L.V., & Olkin, I. (1985). *Statistical methods for meta-analysis*. Orlando, FL: Academic Press.

Hu, T.W., Lee, M.L., & Stromdorfer, E.W. (1971). Economic returns to vocational and comprehensive high school graduates. *Journal of Human Resources*, *6*(1), 23-50.

Jamison, D.T., & Lau, L.J. (1982). *Farmer education and farm efficiency*. Baltimore, MD: Johns Hopkins University Press.

Jamison, D.T., Fletcher, J.D., Suppes, P., & Atkinson, R.C. (1976). Cost and performance of computer-assisted instructión for education of disadvantaged children. In J.T. Froomkin et al. (Eds.), *Education as an industry*. Cambridge, MA: Ballinger.

Lau, L.J. (1970). Educational production functions. In D.M. Windham (Ed.), *Economic dimensions of education*. Washington, DC: National Academy of Education.

Levin, H.M. (1970). A cost-effectiveness analysis of teacher selection. *Journal of Human Resources*, *5*, 24-33.

Levin, H.M. (1974). Measuring efficiency in educational production. *Public Finance Quarterly*, *2*, 3-24.

Levin, H.M. (1976). Concepts of economic efficiency and educational production. In J.T. Froomkin et al. (Eds.), *Education as an industry*. Cambridge, MA: Ballinger.

Levin, H.M. (1983). *Cost-effectiveness: A primer*. Beverly Hills, CA: Sage Publications.

Levin, H.M., Glass, G.V., & Meister, G.R. (1984). *Cost-effectiveness of four educational interventions*. Stanford, CA: Institute for Research on Educational Finance and Governance, Stanford University, Project Report No. 84-A11.

Levin, H.M., Glass, G.V., & Meister, G.R. (1986). The political arithmetic of cost-effectiveness analysis. *Phi Delta Kappan*, *68*(1), 69-72.

Levin, H.M., & Meister, G.R. (1986). Is CAI cost-effective? *Phi Delta Kappan*, *67*(10), 745-749.

Mallar, C.D. et al. (1982). *Evaluation of the economic impact of the job corps program: Third follow-up report*. Princeton, NJ: Mathematica Policy Research.

Manski, C.F. (1985). *Academic ability, earnings, and the decision to become a teacher: Evidence from the national longitudinal study of the high school class of 1972*. Cambridge, MA: National Bureau of Economic Research, Working Paper No. 1539.

Mincer, J. (1979). Human capital and earnings. In Windham, D.M. (Ed.), *Economic dimensions of education*. Washington, DC: National Academy of Education.

Nicholson, W. (1983). *Intermediate microeconomics and its application* (3rd ed.). Chicago: Dryden Press.

Niemiec, R.P., Blackwell, M.C., & Walberg, H.J. (1986). CAI can be doubly effective. *Phi Delta Kappan*, *67*(10), 750-751.

Psacharopoulos, G. (1985). Returns to education: A further international update and implications. *Journal of Human Resources*, *20*(4), 584-604.

Psacharopoulos, G., & Loxley, W. (1985). *Diversified secondary education and development*. Baltimore: Johns Hopkins University Press.

Ragosta, M., Holland, P.W., & Jamison, D.T. (1982). *Computer-assisted instruction and compensatory education: The ETS/LAUSD study, final report*. Princeton, NJ: Educational Testing Service, Project Report #19.

Rao, V.R. (1977). Conjoint measurement in marketing analysis. In J.N. Sheth (Ed.), *Multivariate methods for market and survey research*. Chicago: American Marketing Association.

Ribich, T.I. (1968). *Education and poverty*. Washington, DC: The Brookings Institution.

Sexton, T.R. et al. (1986). Data envelopment analysis: Critique and extensions. In R.H. Silkman (Ed.), *Measuring efficiency: An assessment of data envelopment analysis*. San Francisco: Jossey-Bass.

Smith, A. (1937). *The wealth of nations*. New York: Modern Library.

Spence, M. (1973). Job market signaling. *Quarterly Journal of Economics*, *87*, 355-374.

Stern, D. (1986). Compensation for teachers. *Review of Research in Education*, *13*, 285-316.

U.S. Office of Education. (1970). *Do teachers make a difference*? Washington, DC: U.S. Government Printing Office.

Weisbrod, B.A. (1965). Preventing high school dropouts. In R. Dorfman (Ed.), *Measuring benefits of government investments*. Washington, DC: The Brookings Institution.

Efficiency in Behavior-Changing Social Programs: The Case of In-Home Child Abuse Prevention

Will Johnson, DSW
Tom Clancy, MSW

INTRODUCTION

Many activities that occur as part of government-sponsored social programs involve attempts to change human behavior to make it conform to societal goals and expectations. Examples abound and include such diverse activities as attempts to encourage welfare recipients to find work, remedial education programs, outpatient mental health services to prevent hospitalization, and attempts to rehabilitate juvenile criminals and prevent recurrence of child abuse. Criticism of such programs in recent years indicates that they are widely perceived as inefficient (Morris & Williamson, 1987; AuClaire, 1984; Miller & Gold, 1984). Any employee of a social agency involved in these endeavors who has ever been to a cocktail party can recount numerous stories of incredulous reactions to the news that he or she, seeming otherwise reasonable, would engage in activities thought to be so inefficient in their use of society's resources. Whatever the reasons for such perceptions, they make life difficult for everyone connected with the programs – from clients to workers, administrators, advocates, and politicians.

Dr. Johnson is Supervising Management Analyst, Alameda County Social Services Agency, 401 Broadway, Oakland, CA 94607. Mr. Clancy is Human Services Department Administrative Assistant, Alameda County Social Services Agency.

To the extent that these perceptions are valid, most would agree that policy makers, program administrators and staff are obligated to work to improve efficiency. We believe, however, that many of those who conceive, create, and run such programs, as well as some scholars who study them, may not be aware of what it means to say that a program is efficient or of what efforts would be required to try to improve its efficiency.

To evaluate program efficiency properly, one must understand the meaning of the term "efficiency," and one must know how to determine whether programs are efficient according to that definition. If one then wants to improve efficiency one must understand what additional knowledge is needed and how to get that knowledge. One must also know what forces constrain efficiency and how to overcome them. To elucidate some of the issues and problems involved, the authors will examine the efficiency of a prototypical nationally mandated, publicly funded behavior-changing program – in-home child abuse prevention – and will discuss the implications of the findings for behavior-changing programs in general.

In-home child abuse prevention programs attempt to prevent repetition of abuse and neglect of children referred to children's protective services agencies, without placing the children in foster care. These programs provide an excellent example for analysis because they are perhaps the most important child abuse prevention programs in the United States (Kadushin, 1987), and they possess many of the features typical of behavior-changing programs.

DEFINING AND MEASURING EFFICIENCY

The first task is to clarify what we mean by efficiency. In economics, efficiency is defined as the ratio of outputs to inputs (Seldon & Pennance, 1975). In this standard definition, a cause and effect relationship between the production process (cause) and the output (effect) is implicit. For example, we never question that baked goods (outputs) are caused by (are the effects of) activities of bakers. The physical and temporal proximity of the cause (baking) and the effect (bread), and the lack of other plausible competing expectations for the existence of the bread, overwhelm all doubt that the bread was produced (caused) by the bakers.

Thus, in the numerator of the ratio of outputs to inputs that defines the efficiency of a bakery, we encounter a positive number (number of loaves of bread), and we confidently assume that the bread represented by this number is the result of the efforts of the bakers. This number tells us the bakery does, indeed, possess some base level of efficiency; it is at least somewhat effective at producing bread. Efficiency and effectiveness can be considered synonyms (Guralnick, 1970), though effectiveness is not usually defined as a ratio of outputs to inputs.

For in-home child abuse prevention and other behavior-changing programs, we can also find a positive number to place in the numerator of the efficiency-defining ratio of outputs to inputs. For child abuse prevention programs, this might be the number of cases that have passed through a program without having recurrence of maltreatment subsequently detected.

But, can we take it for granted that the "nonrecurrences" of child abuse represented by this positive number were caused by the program's "production process" (e.g., case management, counseling, etc.) and are thus truly program effects or "outputs"? Can we begin to speak of the numbers of recurrences of child abuse prevented per dollar of program funds spent, just as we would speak of numbers of loaves of bread baked per dollar spent? Not yet.

A long list of plausible alternative explanations for the nonrecurrences of abuse that we observe—other than exposure to the program—must first be eliminated. These include such things as autonomous changes in behavior of program participants due to human growth and development or spontaneous remission from states of disease (maturational processes), the occurrence of events not part of the abuse prevention program that can influence program participants' behavior (historical events), and others (Campbell & Stanley, 1963).

Have these alternative explanations for the "successes" of in-home child abuse prevention programs been eliminated by properly conducted research? Can we speak of the efficiency of in-home child abuse prevention programs in full confidence that they cause nonrecurrence of maltreatment? To answer these questions we must proceed as we would in the case of any behavior-changing program, by reviewing the empirical evidence.

In doing this, we must make our evaluation according to methodological criteria relevant to determining the existence of causal relationships. The necessary research designs were described in detail twenty-five years ago by Campbell and Stanley (1963) in their now classic exposition of requirements for causal inference in the study of human behavior.

According to their well-accepted criteria, the only research designs that allow relatively unambiguous causal inference are true experiments. The key feature of these designs is random assignment of subjects from a common pool to groups that receive different amounts of a treatment or no treatment at all.

In the study of child abuse prevention, a number of investigators have stipulated other requirements that must be met as well. Some of these are a clear description of the maltreatment being studied, a clear definition of the preventive program, and equal detection of outcomes across experimental and control or comparison groups (Leventhal, 1982). Other criteria include reliable measurement of all variables studied and adequate taxonomic delineation of the maltreatment being studied (Besharov, 1982).

Thus, while it is not a small task, the tools are available for properly conducted studies to establish unambiguous cause and effect relationships between receipt of services and prevention of recurrence of maltreatment. In other words, efficiency, as we have defined it, can be measured.

To find studies meeting the criteria discussed above and to assess the evidence that in-home child abuse prevention programs cause nonrecurrence of maltreatment, and therefore possess some degree of efficiency, we performed a brief review of the literature. Some already published reviews offer a convenient way of summarizing and characterizing the evidence.

Howing et al. (1989) reviewed 107 empirical studies. Of these, not a single one having recurrence of child abuse or neglect as the outcome variable met the criteria for demonstration of causation set forth by Campbell and Stanley (1963) and others (Leventhal, 1982; Besharov, 1982). Howing et al. (1989) did cite an experimental study in which the outcome variable was prevention of foster-care placement, as distinct from prevention of recurrence of abuse (Szykula & Fleishman, 1985). In this study, results were equivocal. The authors wrote that the treatment used worked with cases judged "less difficult" (experimental group N = 13, control N = 13) but did not work with cases judged "more difficult" (experimental group N = 11, control N = 11).

In all of the studies reviewed by Howing et al. (1989), lack of control, or comparison, groups was a pervasive problem (e.g., Miller et al., 1984, 1985). Failure to use random assignment to groups in studies that did have control groups was another weakness (Lutzker et al., 1984), and other problems included the use of clearly reactive measures such as caseworkers' ratings of their own success (Daro, 1988). Some seriously flawed research has been republished in so many different forms that assertions of program "effectiveness" contained therein may have begun to appear factual (Daro, 1988).

Howing et al. (1989) ". . . repeatedly called attention to the general lack of systematic evaluation of many intervention modalities currently being used with maltreating parents and their children" and reached only "tentative" conclusions about treatment effects (p. 335).

The results of a review by Blythe (1983) of sixteen studies of child abuse treatment outcomes were similar. Only four of the sixteen contained any measure of recurrence of maltreatment, and only one of these four employed actual observation of the treated subjects in an attempt to find out whether maltreatment had recurred. This lone study employed a single-subject design, thus affording no basis for generalizing to the larger population of cases exposed to child abuse treatment.

Blythe (1983) concluded that "since these studies rarely used child abuse as a dependent variable, little can be said about which treatment packages eliminated or reduced child abuse" (p. 331).

Magura (1981) reviewed studies of the effectiveness of services to prevent foster care placement. He found one study of effectiveness that employed an experimental group and a comparison group to which subjects were randomly assigned (Magura & DeRubeis, 1980). Unfortunately, the finding of this study was that when experimental and control groups were compared there was no significant difference in the rate of foster care placement due to recurrence of maltreatment.

These examples reflect our entire review. There appears to be an absence of studies meeting the requisite methodological standards. Our review, and these examples, also document the equivocal nature of support for the effectiveness of in-home child abuse prevention in the tiny number of studies that do come close to meeting those standards. Simply put, the literature shows no solid evidence that in-home child abuse prevention programs cause reductions in child abuse.

KNOWLEDGE REQUIREMENTS FOR IMPROVING EFFICIENCY

Once one has determined whether a program possesses some base level of efficiency (that is, has a positive ratio of outputs to inputs) using the analytic approach described above, one can then decide to eliminate the program, keep the program as it is, or move on to the task of improving efficiency.

To increase program efficiency above a base level, additional information is needed. If we know that a program works at least to some extent, we must then ask how much service or treatment of a given kind will cause attainment of the maximum probability of a positive outcome for a particular kind of case. This is one possible formulation of what Pruger and Miller (1989) have called the targeting question.

It is clear from this question that in some ways the knowledge we seek in order to increase program efficiency above base levels is more complex than that required to demonstrate an initial base level of efficiency. Studies will have to be more elaborate, to include differing amounts and kinds of treatments and differing samples of cases. In terms of the requirements for establishing causation, however, experiments, as defined and discussed above, must still be performed.

Consider, for example, a study designed to tell whether urban child abusers respond positively to three months of individual counseling combined with parent effectiveness training. In such a situation, it will still be necessary to show that the particular combination of treatments used with the sample of subjects studied caused whatever outcomes are observed, and not other, extraneous events that preceded outcomes (Campbell and Stanley's historical events, maturational processes, etc.).

As just noted, to make such a showing, experiments must still be performed. The only think that changes is the kind of experiment required. For establishing base levels of efficiency, single factor experiments—those that test the effect of a single treatment with a particular kind of client—are the requirement. To improve efficiency, we must test the combined effects of multiple kinds and amounts of treatments with different kinds of clients. Here, multifactorial experiments are required (Keppel, 1973; Campbell & Stanley, 1963).

In such experiments, two, three, or more treatments (program services) would be administered in varying amounts (i.e., dosages) and combinations (e.g., individual counseling plus parent effectiveness training) to differing samples of cases (e.g., urban and rural) assigned randomly to treatment and control groups (Keppel, 1973). Such controlled experiments offer the only way of determining the separate and combined causative roles of the services that are presumed to produce program outcomes.

In our analysis of in-home child abuse and neglect prevention programs, we looked for research of the kind described above. Based on this review, we believe there exist no multifactorial experimental studies that examine the preventive effects of various treatment combinations with different kinds of cases in practice settings. Indeed, we found a lack of evidence establishing that in-home child abuse prevention programs possess even base levels of efficiency. In the absence of such evidence, trying to determine what treatment combinations work with different kinds of cases seems premature.

PUTTING THE INFORMATION TO USE

It is obvious that demonstrating a base level of efficiency and possessing the information to increase efficiency above that level are not enough. Somehow, the information must be put to use to be of value.

In publicly funded behavior-changing programs, both program executives and caseworkers are possible starting points for initiatives to improve efficiency. Executives or managers have direct responsibility for supervision of caseworkers and can decide to attempt to influence case outcomes through their managerial decisions. Caseworkers are also professionals in many jurisdictions, at least nominally, and can take it upon themselves to improve efficiency through improved practice. But, as a look at the situations of these two groups will show, there are forces working against initiatives from either source. Putting information to work in organizations can be at least as difficult as producing the information in the first place.

One well-known problem is endemic to public programs: in the absence of outside pressure, public program managers have few incentives for improving efficiency. In fact, the reverse may be true. For example, funds left unspent can usually not be redeployed for other purposes or to expand program operations. Instead, unused funds are likely to be cut from next year's budget. Thus, even if information that could increase efficiency is available, it may not be used. This is particularly true if use of the information would lead to reducing funding or curtailing program operations.

The workings of administration can be another obstacle to efficiency initiatives from management. For example, staff support to improve efficiency may be lacking, requiring that management rely on techniques of enforcement. Use of these techniques has severe, purely mechanical, limitations (Brodkin, 1986). One is that any enforced attempt to improve efficiency through improved case practice must rely on information about individual case actions taken by individual workers in order to hold them accountable if they fail to perform their duties. But, such information may be impossible to get.

Consider an example from in-home child abuse prevention. Research might conceivably reveal that a much higher level of prevention of recurrence of child abuse (outputs) could be achieved with only a small increase in frequency of worker-client contact (inputs), thus increasing program efficiency.

To enforce this improvement in efficiency without staff support, management would have to have a way of knowing about individual worker's client contacts. Management could not afford to gather this information without extensive computerization not currently present or contemplated in most settings. In California counties, for example, client contact information is kept in handwritten case records. Our analyses of such cases indicate that it takes about 50 minutes per case for a trained rater to read a case record and extract the client contact information for computer analysis. Because of the cost of extracting the information from case records, generating regular reports on child welfare workers' client contacts would absorb most of a county agency's budget and is therefore infeasible.

A final point about restraints on initiatives from management is that vigorous enforcement of regulations implementing improvements in efficiency can have negative effects on labor relations, even threatening the tenure of program executives in extreme cases. Managers wish to prevent the political nightmare created by groups of placard waving, chanting employees descending on local legislators' chambers demanding resignations.

In-home child abuse prevention also offers an example of why large-scale efficiency improvement initiatives are unlikely to come from program staff. There are many possible reasons. More work might be required, workers may feel it is management's responsibility to initiate collective action, etc. But even if workers want to improve efficiency and overcome other restraints, they are hampered by a lack of information about how to do it. Information about how to improve efficiency would come primarily through research reports in the professional literature.

Unfortunately, most public agencies do not treat caseworkers as true professionals. Either because agencies cannot afford to, or for other reasons, staff are not given time to read research reports on the job, and such literature is not generally made available by agencies. So, caseworkers are generally unaware of the contents of professional journals (Freyer et al., 1988). Further, the most widely read journals are unlikely to contain the necessary research reports anyway (Cain & Klerman, 1979).

A final constraint on efficiency improvement initiatives from either managers or workers remains: even if the knowledge necessary to improve efficiency were available, management and staff might resist using it because it would be likely to reduce their autonomy and increase their accountability.

The nature of the work done in behavior-changing programs requires that staff exercise considerable discretion and autonomy in the performance of their duties (Brodkin, 1986). Where social workers are employed, this is reinforced by the norms of the social work profession (Brodkin, 1986). Use of proven general truths based on properly conducted efficiency research would require some standardization of service provision decisions and, therefore, some reduction of staff autonomy and discretion. Our direct experience is consistent with the views of Brodkin (1986) and Lipsky (1980) that staff will not enthusiastically embrace reductions in their autonomy.

And, any attempt to improve efficiency must involve routine measurement of individual and program outputs (numbers and percentages of cases in which recurrence of abuse/neglect was prevented, numbers and percentages of criminals who did not repeat offenses, etc.). For only if one knows one's current level of output can one (or one's supervisor) know whether improvement is needed, and whether it has been achieved. In programs where output measurement is not now routine, this practice will result in increased accountability.

In-home child abuse prevention offers a striking example. We know of no jurisdiction in the United States where the most obvious program outputs – foster care placements prevented and recurrences of abuse prevented – are currently measured and used to evaluate either overall program efficiency or individual case worker efficiency. We think instituting such measurements would, at least initially, encounter a very vigorous negative reaction.

SUMMARY AND DISCUSSION

Publicly funded behavior-changing programs are widely perceived as inefficient. To know whether these perceptions are valid and to improve efficiency where necessary, administrators and policy makers must know what efficiency is, how to gauge the efficiency of programs, what knowledge is needed to improve program efficiency, and how to overcome organizational obstacles to the use of that knowledge.

We have shown that proving the efficiency of a behavior-changing program requires us to establish that the production process used (case management, counseling, etc.) causes whatever "output'' (e.g., abuse prevention) is produced. We have argued that true experiments are the only way to establish causation and are, therefore, the only way to establish base levels of efficiency. We have shown that multifactorial true experiments are required to improve efficiency beyond base levels. And, we have identified organizational forces affecting program managers and line staff that must be overcome in order to improve efficiency once the required knowledge has been generated through research.

In applying the criteria developed in our presentation to the case of in-home child abuse prevention, we found no compelling evidence arguing either for or against this program's efficiency. This is not really surprising; in-home child abuse prevention does not seem to be alone in lacking evidence regarding efficiency. Alcohol abuse prevention and drug abuse prevention are similarly without definitive proof of their efficiency (Moskowitz, 1983, 1989).

The absence of evidence of efficiency does have very serious consequences. While waiting for definitive research, program executives will be unable to provide proof of efficiency to the public or legislators. As a result, funding levels may be lower than they otherwise might be (*Behavior Today*, 1976). Assuming that a program is ultimately found to be efficient, some clients in need of service will suffer from not having recieved it while funding levels are low. For the clients of in-home child abuse prevention, vulnerable abused and/or neglected children, failure to receive effective service is a terrible outcome.

There is another possible outcome that is also very unfortunate. If in-home child abuse prevention is ultimately found to be inefficient (and therefore ineffective), all the resources consumed by the program will have been used unproductively during times when there may be other, more effective solutions, and certainly when there are many other pressing social needs to which the resources could have been allocated.

To avoid experiencing either negative outcome, society must understand the phenomena that produce and maintain the existence of behavior-changing programs even when there is no evidence of efficiency. At their inception, such programs seem to draw their energy from the most basic human impulses. In-home child abuse prevention, for example, may express on society's behalf the ferocious protectiveness that good parents feel when their young are threatened.

Whatever the source of such impulses, when they are unleased, large numbers of people seem to assume immediately that some program (any program?) to address the problem will work (be efficient). The small amount of good evidence on in-home child abuse prevention does not bear out this assumption.

What can be done? The most appropriate starting point would appear to be at the beginning of the process of program development before program beneficiaries have formed constituencies. Useful models might be found outside of social science. For example, to prevent quackery and protect the public health we have laws requiring that all drugs must be shown to be safe and effective (efficient) before they can be sold to the public. The research designs employed in these tests of effectiveness are all extensions of Campbell and Stanley's (1963) true experiments, because medicine has recognized that knowledge of causation cannot be dispensed with.

Some reasonable adaptation of this approach should be tried with publicly funded behavior-changing programs to ensure that they too are efficient—that they do indeed change behavior—before public resources are committed to them. When a program is shown to be efficient, we can wholeheartedly support it, and we can work to make it even better. When a program is shown to be inefficient, we can make an informed decision about what to do next.

REFERENCES

AuClaire, P.A. (1984). Public attitudes toward social welfare expenditures. *Social Work*, *29*, 139-144.

Behavior Today. (1976). Editorial. Quality psychiatric care vs. political realities. *Behavior Today*, *7*(26), 1-2.

Besharov, D.J. (1982). Toward better research on child abuse and neglect: Making definitional issues and explicit methodological concern. *Child Abuse and Neglect*, *5*, 383-390.

Blythe, B.J. (1983). A critique of outcome evaluation in child abuse treatment. *Child Welfare*, *42*(4), 325-335.

Brodkin, E.Z. (1986). *The false promise of administrative reform: Implementing quality control in welfare*. Philadelphia: Temple University Press.

Cain, L.P., & Klerman, L.V. (1979). What do social workers read about child abuse? *Child Welfare*, *38*(1), 13-24.

Campbell, D.T., & Stanley, J.C. (1963). *Experimental and quasi-experimental designs for research*. Chicago: Rand McNally College Publishing.

Daro, D. (1988). *Confronting child abuse: Research for effective program design*. New York: Free Press.

Freyer, G.E. Jr., Poland, J.E., Bross, D.C., & Krugman, R.D. (1988). The child protective service worker: A profile of needs, attitudes, and utilization of professional resources. *Child Abuse and Neglect, 12*, 481-490.

Guralnick, J. (Ed.). (1970). *Webster's new world dictionary of the American language*. 2nd College Edition. New York: The World Publishing Company.

Howing, P.T., Wodarski, J.S., Gaudin, J.M. Jr., & Kurtz, P.D. (1989). Effective interventions to ameliorate the incidence of child maltreatment: The empirical base. *Social Work, 34*(4), 330-338.

Kadushin, A. (1987). Child welfare services. In *Encyclopedia of Social Work*, Vol. 1, 18th Edition. Silver Spring, MD: National Association of Social Workers.

Keppel, G. (1973). *Design and analysis: A researcher's handbook*. Englewood Cliffs, NJ: Prentice-Hall, Inc.

Leventhal, J.M. (1982). Research strategies and methodologic standards in studies of risk factors for child abuse. *Child Abuse and Neglect, 6*, 113-123.

Lipsky, M. (1980). *Street level bureaucracy: The dilemmas of individuals in public services*. New York: Russell Sage.

Lutzker, J.R., Wesch, D., & Rice, J.M. (1984). A review of project "12-ways": An ecobehavioral approach to the treatment and prevention of child abuse and neglect. *Advances in Behavior Research and Therapy, 6*, 63-73.

Magura, S. (1981). Are services to prevent foster care effective? *Children and Youth Services Review, 3*, 193-212.

Magura, S., & DeRubeis, R. (1980). *The effectiveness of preventive services for families with abused, neglected and disturbed children: Second year evaluation of the Hudson County Project*. Trenton, NJ: Bureau of Research, Division of Youth and Family Services.

Miller, K., Fein, E., Howe, G.W., Gaudio, C.P., & Bishop, G.V. (1984). Time-limited, goal-focused parent aide service. *Social Casework, 65*, 472-477.

Miller, K., Fein, E., Howe, G.W., Gaudio, C.P., & Bishop, G.V. (1985). A parent aide program: Record keeping, outcomes and costs. *Child Welfare, 44*, 407-419.

Miller, M., & Gold, M. (1984). Iatrogenesis in the juvenile justice system. *Youth and Society, 16*(1), 83-111.

Morris, M., & Williamson, J.B. (1987). Workfare: The poverty dependence trade-off. *Social Policy, 18*(1), 13-16, 49-50.

Moskowitz, J. (1983). Preventing adolescent substance abuse through drug education. In T. Glynn, C. Leukefeld, & J. Ludford (Eds.), *Preventing adolescent drug abuse: Intervention strategies*. DHSS Publication No. (ADM) 83-1280, 233-249.

Moskowitz, J. (1989). The primary prevention of alcohol problems: A critical review of the research literature. *Journal of Studies on Alcohol, 50*, 54-88.

Pruger, R., & Miller, L. (1990). Efficiency in the social services: Part 2. *Administration in Social Work*, *15*(1 & 2).
Seldon, A., & Pennance, F.G. (Eds.). 1975. *Everyman's dictionary of economics*. London: I.M. Dent & Sons.
Szykula, S.A., & Fleischman, M.J. (1985). Reducing out-of-home placements of abused children: Two controlled field studies. *Child Abuse and Neglect*, *9*, 277-283.

PART 3: EFFICIENCY IN THE AGENCY AND THE CURRICULUM

Efficiency Considerations in the Social Welfare Agency

Joseph Heffernan, PhD

This volume is concerned with the pursuit of efficiency in the provision of social services. This paper is concerned with the normative concepts of efficiency and effectiveness and a description of one quantitative method of measurement, Data Envelopment Analysis, which has some potential to treat the critical issues in a less acrimonious way.

Efficiency considerations have always been prominent in the minds of administrators concerned with achieving the most they can from available resources. At the same time, they often harbor a sense of hostility toward efficiency notions as well as innovations made in the name of efficiency. Professional social workers on the line, and other staff members, are often in conflict with administrators over the right place for the use of efficiency criteria. The con-

Dr. Heffernan is Professor, LBJ School of Public Affairs and the School of Social Work, The University of Texas at Austin, Austin, TX 78712.

flict is perhaps stronger in the social service field because of a connotation that efficiency considerations are linked to market considerations and appear to conflict with the ideal of service, altruism and caring. This ambivalence about the connotative aspects of efficiency is revealed in informal conversations with staff members, supervisors, bureau chiefs and agency directors, and political executive officers. Both those concerned directly with practice and those concerned with administrative aspects of the agency have a commitment to efficiency, but both groups have a sense of unease about how and under which conditions an efficiency consideration should dominate the choice.

The participants in an internal agency policy debate often appear to be talking past one another (Stone, 1980). There are at least three streams of reference. One concerns efficiency and managerial control. At this level, the problem is perceived as a problem over patterns of allocation of fixed resources among competing ends. The second focus is fixed on notions of effectiveness. The question here is the essentially technical problem of selecting the best way to use fixed resources to accomplish a single well-specified end. The third focus is on the necessity to secure a consensus among participants at all levels within the organization. The internal conflicts between professional staffs with their concern with effectiveness and the efficiency criterion of administrators are legitimized by participants making reference to one of these three criteria (Leathers, 1979).

Pressed to define efficiency, most social workers provide a close approximation of some text book definition of technical efficiency and make a clear point that efficient and least expensive are not identical. Discussion of the efficiency within their own previews of their own jobs reflects a serious concern with measurement problems.

In the professional and academic literature, three streams of speculation merge in discussions about efficiency in social work. One has to do with the need to identify better techniques for the observation and measurement of clinical outcomes (Rubin, 1985). The second has to do with the way in which evaluative and implementation studies are incorporated into political conflicts over objectives (Kelman, 1987; Stone, 1980). The third addresses the epistemological issue of the utility of the efficiency criteria outside of market conditions (Cleveland, 1979; Kelly, 1980). These three issues also dominate the pragmatic thoughts of agency administrators as they try to use efficiency criteria in their own decision making processes.

Within the public or voluntary social agency, the criterion most frequently employed to structure allocation decisions is the mission perspective. Section chiefs and professional staffers want to have their accomplishments linked to statements of objectives, strategic plans, and other devices which, while somewhat less objective, can be a functional substitute for the profit and loss statement. The efficiency criterion is based on net return to investment and assumes that all appropriate costs and benefits can be tabulated. The efficiency criterion further assumes that, under conditions of *ceritus parabus*, efficiency is an unambiguous good. Both agency executives and staff professionals tend to reject this hard line notion. Clearly, there are many instances when either society or an agency find it desirable to reject efficiency gains on ethical and/or equity grounds. While the efficiency criterion is sometimes not appropriate, this does not mean it can be ignored.

Although the empirical research base is weak, we do know from case studies and practical experience that many factors influence the content of decisions about the productive processes and allocation patterns within each agency. Efficiency considerations and effectiveness considerations are generally thought to have a special sort of prescriptive utility. Yet, efficiency and effectiveness are sometimes seen as counter to one another in the non-profit sector. Social policy instructors often assert that such beliefs indicate a lack of conceptual clarity (Knapp, 1984). A more pragmatic perspective emerges from conversations with administrators. It is that the imprecise and ambiguous use of "efficiency" and "effectiveness" is a deliberate result of conflict over agency objectives. The terms are more than instruments of analysis; they are also weapons in political conflict. Agency administrators appear to be aware of this in their own thinking and are acutely aware of it when suggested measures of (or opposition to) efficiency are provided by subordinates. The professor in the classroom may have the power to enforce a particular definition, but in the real world of agency decision making such clarity of terms is difficult to maintain (Shubick, 1978).

Consistent use of normative modifiers attached to the term "efficiency" is one way that a mutually dependent and consistent set of perspectives with regard to the place and value of efficiency can be formed. Below is one attempt at such clarity:

Allocative efficiency: This is the broadest concept in which all of the decisional rules below are merged. The general allocative efficiency criteria is the Pareto principle which is achieved when no one impacted by the decision can be made better off without someone else being made worse off. While conceptually the most clean, it is the least useful in practice.

Cost efficiency: A procedure where all the benefits and all of the costs are measured by a single numeraire—normally the dollar. Comparisons between option is made by reference to one of three "decisional rules":

1. Select the option with the highest present value and reject all programs with negative present values.
2. Select the option with the highest rate of return and reject any option with a rate of return less than the prevailing discount rate.
3. Select the option with the highest benefit cost ratio and reject all options with benefit cost ratio less than one.

Instrumental efficiency: This view places its emphasis on the manipulation of inputs to achieve given objectives. Viewed in this way, that which is important for the decision is not so much how the output is to be valued but the various combinations of resource use and productive processes which increase the probability that particular outcomes will be achieved. Comparisons between option is made by reference to one of two "decisional rules":

1. Select the option with the highest "gap effectiveness," i.e., the degree to which the problem is abated.
2. Select the combination of options that is least costly for a given end.

Administrative efficiency: This is not only a measure of the relationship of administrative costs to the agencies total operation. It is also a measure of how well those administrative costs are utilized to target the benefits and the services of the agency. Comparisons between option is made by reference to one of three "decisional rules":

1. Select the option which minimizes the ratio of administrative costs to service costs.
2. Select the option that maximizes the probability that all of those intended to be served are actually served (inclusive efficiency).
3. Select the option that maximizes the probability that only those who are intended to be served are served (exclusive efficiency).

MEASURING EFFICIENCY IN THE AGENCY: TOWARD MATHEMATICAL MODELS

A clear implication of the framework set out above is that we measure efficiency and judge efficiency in a variety of contexts. Efficiency is a comparative concept. Calculating inputs and outputs will not provide a clear prescriptive rule for an agency's decisions about allocation patterns without some yardsticks against which the calculations are compared. In general, three distinct kinds of comparisons are made: (1) the agency is measured against its own past performance; (2) the agency is measured against essentially similar agencies; (3) the agency is measured against some theoretical standard which specifies a form of ideal performance.

Efficiency to the economist or the engineer refers to the relationship between outputs and inputs. Basic to this concept is the notion of allowance for factors outside of the control of the actor or agency. The economist assumes that efficiency is evaluated under condition of *ceritus parabus*, other things being equal. Social workers too often fail to consider this basic assumption. The trade off between efficiency and equality requires complex Pareto assumptions. At any level of effectiveness, efficiency is achieved by increasing the output from given inputs or reducing the inputs required for a constant output. Under conditions of scarcity, and *ceritus parabus*, efficiency is always desired. The important thing is to focus on the context within which the concept is evaluated. It is important to point out that Arthur Okun's trade off was not between efficiency and effectiveness but rather between efficiency and equality (Rimer, this volume; Okun, 1975). The critical need is to locate common definitions and common indicators. This may not have the overriding prescriptive power that economist claim for it; but it does hold some promise of clarifying the debate.

DATA ENVELOPMENT ANALYSIS AND RATIO EFFICIENCY ANALYSIS

One procedure for the measurement of instrumental efficiency in non-profit contexts is known as Data Envelopment Analysis (Banker, 1989). In DEA analysis, neither the output nor the input needs to be measured in dollar terms. In a subclass of DEA, known as Ratio Efficiency Analysis, a procedure fits mathematical programs to observed data in order to locate a production function which results from using specific inputs. This allows the investigator to make relative efficiency judgments. In DEA, the entities responsible for committing inputs in pursuit of outputs are called decision making units (DMU's). Depending on how the data are aggregated, these may be individual social workers, agencies, or divisions of larger bureaucracies. All that is required is that the DMU's to be compared pursue at least one common objective and use at least some common resources in pursuit of that objective. The model can be illustrated by reference to a scenario of a set of agencies seeking to accomplish family restorations by using foster care, homemakers, and day care.

In this scenario

$$FR = (X)FC + (Y)HM + (Z)DC$$

where FC is foster care
HM is homemaker service
DC is day care

and X, Y, and Z represent the number of units that are required to achieve the goal of family reunification (FR).

Generalizing from this model, one could specify that:

$$FR = F(a\ FC,\ b\ HM,\ c\ DC) + e$$

With a, b, and c as the coefficients for an exclusive and exhaustive list of resources used, e becomes the error term. The success of the model is reflected in the size of the coefficient of determination (R 2). Using exemplary data from the table, the regression line becomes an average production function. Output is a function of a set of inputs. The vertical deviations from average performance is measured by the standardized residuals. The plus or minus sign of each agency's standardized residual is an indicator of the efficiency. The agencies with a positive sign are efficient while those with a negative sign are inefficient. The magnitude of the standard residual is a measure of relative efficiency. The results of the estimation give the standardized residuals that are listed in column six of Table 1.

This measure of efficiency will be only loosely correlated with a cost/benefit ratio since the DEA method does not attach shadow prices or other estimates to the cost of either inputs or outputs. What DEA reveals is the instrumental efficiency with which available resources are put together to accomplish agency objectives.

When there are sufficient observations, the procedure also allows for the evaluation of the relative performance of each DMU with each resource used. When there is a single input being evaluated, the solid line in Figures 1, 2, and 3 becomes an estimate of the production function for that input. DMU's above and to the left of the production line are efficient while those below and to the right are relatively inefficient. More importantly, each DMU can be ranked relative to all others without regard to budget constraints by examining the sign and magnitude of the standardized residuals. This information for our exemplary data are given in Table 2.

TABLE ONE

DMU Outputs as a Function of Fixed Inputs

	Agency	FR's*	Homemaker	Day Care	Foster Care	Instrumental Efficiency DMU Std. Res
1	A	19	3	5	2	1.069
2	B	25	3	6	4	1.534
3	C	23	3	13	1	1.516
4	D	17	2	5	1	1.624
5	E	16	3	1	2	0.862
6	F	15	3	6	1	0.102
7	G	24	3	11	3	0.774
8	H	17	3	5	2	0.304
9	I	21	2	7	4	0.292
10	J	18	2	4	3	0.647
11	K	22	5	0	5	-0.044
12	L	15	0	9	2	0.166
13	M	19	1	16	1	0.325
14	N	16	0	14	1	1.170
15	O	14	1	10	1	-0.176
16	P	16	3	12	2	-1.723
17	Q	13	2	3	2	-0.233
18	R	10	2	3	1	-0.583
19	S	15	0	9	3	-0.631
20	T	12	3	2	2	-0.903
21	U	12	2	3	2	-0.616
22	V	11	3	1	2	-1.050
23	W	13	0	9	2	-0.599
24	Y	20	5	9	3	-1.330
25	X	14	2	10	2	-1.496

*Family Restorations

CONCLUSION

Pruger and Miller have suggested in this volume that "Efficiency is both a conceptual tool and a vision of perfection in the social welfare enterprise; however, the tool is not used and the vision is neither understood nor valued." Case studies, simple memory and conversations about efficiency indicate that efficiency as a tool and as a concept is not so much ignored as it is converted into a political weapon. This political obstacle further confounds the already existing difficulties of establishing valid observations, reliable measurements and a consensus on what dollar value to use. The DEA models remove the problem of dollar evaluation and focus attention

FIGURE ONE

Restorations by Homemaker Units

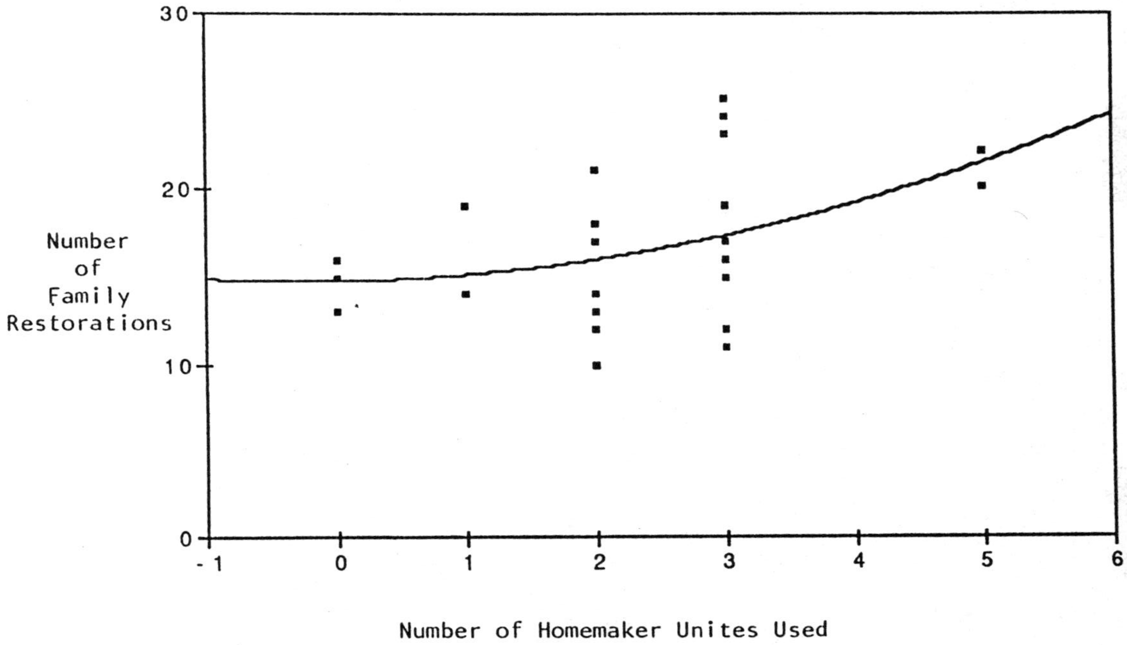

FIGURE TWO

Restorations by Day Care Units

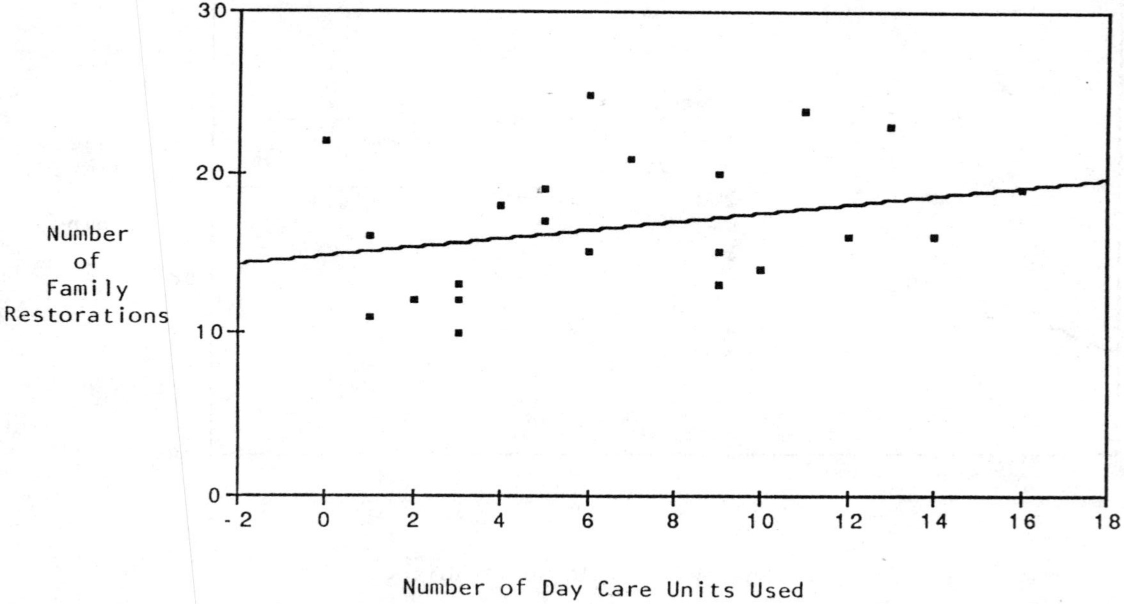

FIGURE THREE

Restorations by Foster Care Units

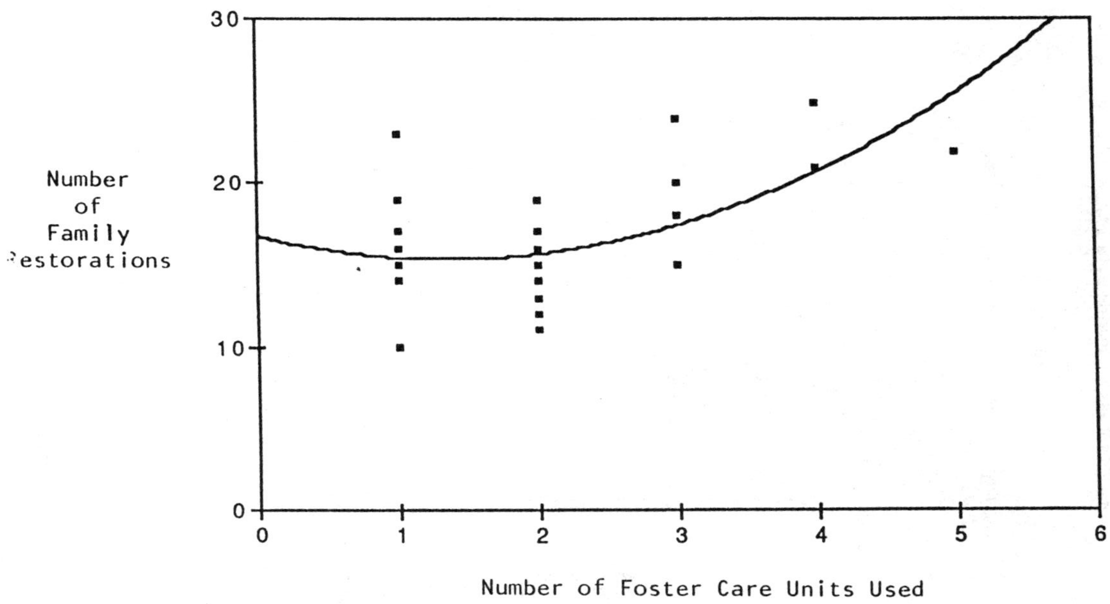

TABLE TWO

DMU Outputs as a Function of Fixed Inputs, Ranked by Instrumental Efficiency

	Agency	FR's*	Homemaker	Day Care	Foster Care	Instrumental Efficiency DMU Std. Res
1	D	17	2	5	1	1.624
2	B	25	3	6	4	1.534
3	C	23	3	13	1	1.516
4	A	19	3	5	2	1.069
5	E	16	3	1	2	0.862
6	G	24	3	11	3	0.774
7	J	18	2	4	3	0.647
8	M	19	1	16	1	0.325
9	H	17	3	5	2	0.304
10	I	21	2	7	4	0.292
11	N	16	0	14	1	0.170
12	L	15	0	9	2	0.166
13	F	15	3	6	1	0.102
14	K	22	5	0	5	-0.044
15	O	14	1	10	1	-0.176
16	Q	13	2	3	2	-0.233
17	R	10	2	3	1	-0.583
18	W	13	0	9	2	-0.599
19	U	12	2	3	2	-0.616
20	S	15	0	9	3	-0.631
21	T	12	3	2	2	-0.903
22	V	11	3	1	2	-1.050
23	Y	20	5	9	3	-1.330
24	X	14	2	10	2	-1.496
25	P	16	3	12	2	-1.723

*Family Restorations

on the critical problem of observation and measurement of comparative performances.

REFERENCES

Banker, R. et al. (1989). An introduction to data envelopment with some of its models and their use. *Governmental and Non-Profit Accounting*, *5*, 125-163.

Cleveland, V. (1979). The meaning of efficiency. In T. Geiger (Ed.), *Welfare and efficiency*. New York: Macmillan.

Kelly, R. (1988). *Promoting productivity in the public sector*. New York: St. Martins Press.

Kelman, S. (1987). *Making public policy*. New York: Basic Books.

Knapp, M. (1984). *The economics of social care*. Oxford: Martin Robertson.
Leathers, C.G. (1979). Language barriers in public productivity analysis: The case of efficiency and effectiveness. *Public Productivity Review, 3*.
Okun, A. (1975). *Equality and efficiency: The big tradeoff*. Washington, DC: The Brookings Institution.
Pruger, R., & Miller, L. Efficiency and the social services, Part A. *Administration in Social Work, 15*(1/2).
Rimer, E. (1990). The impact of efficiency on social work administration education. *Administration in Social Work, 15*(1/2).
Rubin, A. (1985). Practice effectiveness: More grounds for optimism. *Social Work, 30*, 469-476.
Shubick, M. (1978). On the concepts of efficiency. *Policy Sciences, 9*, 121-26.
Stone, C. (1980). The implementation of social programs: Two perspectives. *Journal of Social Issues, 36*.

The Impact of Efficiency on Social Work Administration Education

Edward Rimer, DPA

The failure to understand the meaning of efficiency within the social work profession is also evident in the writings on social work administration education. In fact, social work administration education appears to have developed without any major philosophical foundation. Patti's (1987) argument that effectiveness should be the primary criterion to evaluate the management of social service organizations is the first attempt to provide any type of comprehensive framework to this topic. Patti, however, fails to understand the relationship between efficiency and effectiveness when he notes that efficiency should be "subservient" to the goal of effectiveness. He further reflects that not managing for service effectiveness "explains why management has never been fully understood or accepted in social work and other human service professions." This article is written from the perspective that the lack of understanding, acceptance, and utilization of efficiency as appropriate criteria in social work administration has contributed to the present less-than-satisfactory status of management within the social work profession.

Social work administration education literature tends to evidence the same errors in understanding the relationship between effectiveness and efficiency or it ignores the concept of efficiency entirely. Within the latter category is the endless debate about the relationship between clinical and administrative skills, competition with other administration degree granting programs, and which administrative positions are being filled by MSW graduates. (The entire

---

Dr. Rimer is Assistant Professor, Jane Addams College of Social Work, University of Illinois at Chicago, Chicago, IL 60680.

issue of *Administration in Social Work*, 11(2), Summer 1987, is devoted to social work administration education; also see Gummer & Edwards, 1988.)

Misconceptions regarding the meaning of efficiency for social workers are most pronounced when attempts are made to identify the appropriate administrative skills necessary for social service administrators. For example, Hart (1988) begins her otherwise quite cogent and forceful argument for the inclusion of quantitative management in social work administration curriculum by stating:

> The complex relationship between efficiency and effectiveness has long been explored in the policy and social administration literatures. In 1975, for example, Arthur Okun, delineated the implicit tradeoff between efficiency and effectiveness. Much of the social service literature since that time has appreciated the value of effectiveness as the independent variable and efficiency as the dependent (Turem, 1986). And the business literature has been focused the other way around. The purpose of this paper is to look at the relationship between business skills as tools for efficient management, and leadership as a set of skills for effective management. (p. 1)

Effective and efficient management are viewed as discrete components that require quite different administrative skills. The relationship between efficiency and effectiveness is overlooked, and the preference for effectiveness over efficiency is clearly implied as ". . . an appreciation of technical business skills [efficiency] can be used to augment the leadership repertoire [effectiveness]" (Hart, p. 5). To make matters worse, effectiveness is also used inappropriately. Okun's (1975) work deals with the conflict between efficiency and *equality*, not effectiveness. While both equality and effectiveness are laudable values for administrators, they are by no means either identical or interchangeable.

Given that there is a widespread misunderstanding of the efficiency concept and that presently social work administration education lacks a well defined and agreed upon focus (Patti, 1985), this analysis will center on the present structure of social work administration education. The value system that supports our present curriculum will also be examined.

THE STRUCTURE OF SOCIAL WORK ADMINISTRATION EDUCATION

The present debate on the structure of administration education is framed around the issue of integration versus separation (Neugeboren, 1987). Questions about what, if any, administration content should clinical students be exposed to, and conversely, to what extent should clinical content be required of administration students emphasize the structural dilemmas that need to be resolved *within* existing social work programs. The extent to which clinical and administrative students are exposed to the same content reflects curriculum integration. The structural issue is also center stage when considering the relationship of the MSW to the MBA, MPA, MPH, and other administration degree granting programs. In this case the question of integration is external to the program, in fact external to the profession itself. What is at issue here is program integration.

The present debate on these issues has taken place in a vacuum that rarely includes any concern for issues of efficiency on at least two important levels. First, when identifying skills necessary for clinical and administrative social work (the starting point for discussions on curriculum integration), the importance of efficiency as a goal to be achieved in both clinical and administrative social work is conspicuous only by its absence.

Curriculum Integration

Social workers have historically complained that their caseloads are excessive. While this may often be correct, they overlook the more obvious fact that there is only a limited amount of time in the day, week, or month in which to do the job. The nature of social work is such that there is always something that should be done but is not because of insufficient time. A focus on efficiency would have students as prospective social workers examine how best to use their time, which clients should be served, how much time should be allocated to each family, child, etc. A common phenomenon in social service agencies (though never statistically proven) is that clients whose surnames come at the beginning of the alphabet receive greater amounts of service. This occurs simply because when a new worker takes over an existing caseload and begins to read the case records which are usually alphabetized, services are provided to those clients, and the worker never has the time to pursue other cases unless emergencies arise. Workers are not trained to evaluate systematically the needs of their entire caseload and to develop realistic service plans and goals, given the limited amount of time that they have to perform their duties. Even if they were, supervisors are ill-equipped to review such actions of workers.

Social work administrators, too, face time constraints that prevent them from doing all that is necessary. Examples include grant proposals that are not written, community meetings that go unattended, or written memos rather than face to face contact. Not all staff receive an adequate amount of supervision. Just as there will always be clients who can benefit from additional service, there will always be workers in need of additional supervisory guidance who do not receive it because the administrators' attention is focused elsewhere. Resource constraints faced by both workers and administrators are well-known and omnipresent. It should be self-evident that knowing how to allocate scarce resources, whether child care funds to clients (by workers) or personnel to different programs (by administrators), has a direct impact on program effectiveness. Efficiency as a goal for social workers and administrators means teaching not merely how to perform certain functions, but when such activities should take place and how to identify who is the client, worker, or program most able to benefit.

Participation and Efficiency

In some situations, the incorporation of efficiency will also have implications for how certain administrative tasks are to be performed. An examination of the renewed interest in the appropriateness of shared decision making in social service agencies (Packard, 1989; Malka, 1989; Hooyman, Fredriksen, & Perlmutter, 1988; Gummer, 1988; Russell et al., 1984) reveals both how the concept of efficiency is ignored in social work administration education and how the inclusion of efficiency would alter administrative practice.

This latest round of research finds that participatory decision making is not as prevalent as expected. Patti (1983) noted that "social welfare managers probably use a participatory strategy more consistently and on a broader range of issues than is true of their counterparts in other types of organizations" (p. 31). Representative of the latest findings, Malka (1989) concluded that despite the fact that participatory management "enhances worker satisfaction and significantly reduces absenteeism on the job . . . managerial behavior in most human service agencies is less participative and is less likely to utilize participation" (p. 61). A study of social work supervisors performed by Russell et al. (1984) revealed minimal adherence to a "team building" administrative style, a method closely associated with participation.

Patti's mistaken belief, and one no doubt shared by others, is based on the premise that the primary obstacle to participative management is ideological, and that social service administrators operating within the prevailing values of the social work profession would readily accept an administrative posture that empowers others in the organization. Assuming that Patti's assumption is accurate about the belief system of social work administrators, we are forced to look elsewhere for an explanation for the lack of group decision making. It is also probably safe to discard as a reason that administrators lack the ability to implement a participative system since it requires strong interpersonal skills, a common ingredient in social work practice.

We are left with only one reason which vividly demonstrates the lack of interest in or understanding of the importance of efficiency: participative decision making is too time-consuming. The arguments for participation rest on traditional notions of improved performance and job satisfaction; however, there is no mention of the potential efficiency losses (and, therefore, reduced effectiveness) that accrue when decisions are routinely processed through large groups of people. Experiences with group decision making often result in the statement, "This meeting was a waste of time." Similarly, proponents of participatory management ignore the efficiency gains that can be achieved by having those with the most knowledge participate.

A recent study conducted on a social service agency that uses a participative decision making structure bore this out conclusively (Rimer, 1989). Participation works when not all decisions are shared in the same manner. Participation is not a worthwhile goal merely because of the process. It is most effective when those with specific knowledge and skills are involved in issues that require their expertise. The top administrators of the agency studied have in essence developed a model of delegation that is based upon efficiency criteria: involve those who have the most knowledge. The system recognizes each individual's "zone of indifference" and seeks to maximize an individual's contribution to the organization (Barnard, 1938).

Participation, therefore, should be advocated as a means of advancing administrative efficiency. It is the lack of efficient implementation that prevents its more widespread use and prevents the sought after gains in improved job satisfaction and organizational performance from being realized.

Other functions of the social worker as administrator and clinician would be similarly affected. Curriculum integration can be enhanced as students in both the clinical and administration tracks focus on efficiency as a desired outcome of their actions. It is conceivable that when workers and administrators comprehend the basic similarity of their jobs, i.e., trying to do more with limited resources or attempting to maintain the same level of performance with diminishing resources, some of the stress associated with "role transition," believed to be a cause of poor administrative performance, will be reduced (Scurfield, 1980).

Resistance to an educational approach that emphasizes efficiency as the desired outcome of both clinical and administrative practice is likely to be strong, and will come from both colleagues and students. While the arguments contained elsewhere in this volume may be convincing on an intellectual level, implementation would require that either current programs be lengthened or that some present material be eliminated, actions not likely to follow quickly from mere intellectual agreement. Learning about efficiency, it will be argued, would divert attention away from learning how to provide services or perform administrative duties and is, therefore, inefficient.

These arguments can be countered in two ways. One, students' field placements can be utilized to emphasize the importance of efficiency. It is in these real life situations where students must confront clients with the limited skills and resources at their disposal that the issue of efficiency becomes central to the provision of services. Field supervisors, who must deal with this issue on a daily basis would make ideal and potent allies in the struggle to convince students of the importance of efficiency. Research directed at identifying and solving the problems of allocating scarce services to clients would convince practicing social work professionals of the importance of this effort. A second avenue is to begin to identify ways in which the present curriculum can be made more efficient. This would result in maintaining those aspects of the present curriculum that we do best and identifying those other curriculum areas that are better supplied by other professional disciplines. Utilizing other professional disciplines is considered program integration.

Program Integration

Faherty's (1987) analysis of core content areas of a sample of MSW, MPA, MPH, and MBA programs indicates that some of the MSW content is duplicated in each of the other degree programs in at least one specific area. A cursory look at any of the established texts in social work administration and human service organizations (Hansenfeld, 1983; Weiner, 1982) would support this conclusion, as the authors of these textbooks draw from standard works in health, public, and business administration. Faherty also found that MSW programs were comparatively weak in the area of math, economics, finance and accounting management. Given that within the social work profession the concept of efficiency is at best misunderstood, the best way to introduce efficiency would be through other academic departments. For most social work faculty to teach about efficiency is equivalent to hiring the fox to guard the hen house. A required course in micro economics for all social work students prior to admission, in much the same manner that many schools require an elementary statistics course for research, would ensure that students have a correct understanding of efficiency. Efficiency as a goal could then be a component of all practice coursework, clinical and administrative, just as we now identify specific practice issues related to gender and minorities.

Practice courses would then focus on identifying what client can benefit most from our intervention, what program will lead to the greatest positive change in client situation, or how certain administrative functions ought to be carried out to enhance organizational efficiency. Some writers have already identified the appropriate techniques that can assist us in this effort. Hart (1988) notes that operations management techniques can be used to understand capacity problems and that:

> Concepts and techniques of cost accounting offer another set of possible efficiency tools to the social administrator. Here is an applied science that differentiates among types of costs such as historical costs, standard costs, overhead, variable, differential, marginal, direct, and opportunity costs. The utility for the social services of such definitional fine-tuning is in its flexibility. (p. 6)

Budgeting is another managerial technique that can provide data necessary to promote efficiency.

> Choices regarding the relative value of different activities can be made once current use of staff time has been calculated and analyzed. (Hart, 1988, p. 7)

Instruction in these techniques and their application to the social services is best left to the social work community. However, it will require that we develop the teaching materials (primarily case studies) that are firmly grounded in social service organizations. Blostein's research (1985) revealed that in those few social work programs that include finance and accounting in their curricula the materials used are often from the non-profit and public administration fields.

Utilizing existing courses from other programs and adopting techniques from other professions is not simply the best way to learn about efficiency or the most efficient way to do so. There is a need for social work students to interact with public, non-profit, and even business administration students, as social workers may find themselves working in various types of organizations. The eternal turbulence of social service delivery makes it difficult to predict in what organizational realm our students will locate jobs. By limiting their exposure to social work organizations and by ignoring the fact that other types of organizations play a significant role in the deliver of social services, we seriously limit the ability of our students to become effective administrators.

An increase in the use of courses and concepts from other disciplines, however, will require a re-examination of the value system that has led to the present social work administration curriculum.

SOCIAL WORK ADMINISTRATION: THE VALUE DILEMMA

If efficiency criteria are to be included as a goal of social work administration education as we outlined above, two issues must be re-evaluated. One is the belief that there is a distinction between micro and macro practice based upon different sets of skills and values. The other strongly held belief is that social service organizations are distinctly different from other types of organizations and that techniques used in other organizations and based upon a discipline other than social work should be suspect (Patti, 1982).

Of major concern is to what extent the values of social work administrators are congruent with those of line social workers (Scurfield, 1980). This issue has led Patti (1985) to state:

> Service effectiveness is most fully congruent with the values and purposes of the human service professions. Administration has never been fully understood and accepted in social welfare, in part, I suspect, because the goals of this method have been perceived to be in conflict with, or only tangentially related to, the field's main business of changing people and social conditions. (pp. 3-4)

Administration has been misunderstood because the relationship between efficiency and effectiveness has been misunderstood. This is evident when Patti (1987) notes:

> An approach to management that seeks to maximize service effectiveness must deal with how this performance criterion interacts with other kinds of organizational outcomes such as output and efficiency . . . . Indeed, it would seem to follow that as the manager diverts resources to improve service effectiveness, at some point this will begin to negatively effect output and/or efficiency. (pp. 13, 14)

When a manager allocates funds to inefficient activities, overall service effectiveness is reduced because those funds are not used where they can do the most good. One cannot be concerned about service effectiveness without also being concerned about efficiency, because social work administrators do not have unlimited resources.

Efficiency is defined as either doing more with the same amount of resources or producing the same output with less. To meet the conditions of the first situation, an organization clearly must be concerned with effectiveness for that is what is being improved. Efficiency means good administration, and good administration is not contrary to either the values or practice of social work (Weinbach, 1990). Efficiency is not only the central purpose of administrative behavior, but is also a force that defines the actions of line social workers, whether they know it or not.

Inherent in social work practice is a belief in change, that clients can improve their life situation, that society will assume more responsibility for those least able to care for themselves. Social workers move into administrative positions for the change to make a positive change in how the organization is functioning ("I can do it better") and not merely for the increased financial remuneration. Social workers interact daily with clients and political, social, and economic systems to increase the potential of their clients, to maximize the social output. Social workers are continuously coping with limited resources, whether those of their clients (poor work skills, AFDC income) or a social system with inadequate schools and housing, trying to extract the maximum gain, to bring about the most positive change in the client situation. This is the accepted, in fact, the expected norm of social work.

When social workers are given too many cases, professional ethics do not permit them to do less than needed with each client. Workers do not reduce their level of expectations of clients or of themselves. They attempt to accomplish the same goals with less time. Social worker burnout is as much a cause of failed attempts at improved efficiency as anything else.

Another reason why administration has not been properly conceptualized within the field of social work and why efficiency has been either misunderstood or ignored emanates from the belief that social service organizations are unique entities.

The Council on Social Work Education defines administration as a "field of practice." Administration is considered to be social work in an organizational, or administrative setting. An alternative perspective might hold that it is administration that occurs in a social work setting.

One consequence of CSWE's definition is that schools of social work develop administration components rather than seek linkages with other professional administration degree programs. Sixty-two percent of the accredited social work programs have a specialization in administration (CSWE, 1987); however, only 26% of these schools have developed dual or joint degree programs in this area. Therefore, only 16% of all the accredited schools of social work have dual or joint degree programs in the area of administration. While there are, no doubt, some schools of social work located in colleges and universities without other administrative degree programs, far more schools of social work have the opportunity to collaborate with other administrative degree programs but choose to ignore them.

The CSWE requirement that only faculty with MSW's teach practice courses results in limits being placed on the content of administration courses. Since it has been only during the past 25 years that schools of social work have offered specializations in administration, those currently on the faculty with MSW's tend to have limited formalized administration training.

By considering administration a field of practice (similar to practice with children, adults, or families), an emphasis is placed on those techniques that are established within the individual social work model, for example, interpersonal communication (Wolk,

Way, & Bleeke, 1982). Efficiency, misunderstood by clinicians is, therefore, ignored as an integral part of administrative training.

Social work administration is administration first. That it takes place in a social service setting is of secondary importance when identifying ways to improve organizational performance. "Borrowing skills and knowledge from other professions is a long-standing tradition in social work" (Hart, 1989, p. 9). Efficiency has its place in social work education for clinicians and administrators.

SUMMARY

Pruger and Miller note that "social workers at all hierarchical levels make discretionary decisions about how scarce resources are used. Efficiency can frame and inform these choices" (Efficiency, Part B, this volume). An immediate impact of efficiency on social work administration education would be to foster greater integration between clinical and administrative students. It is also conceivable that it would bring about a closer relationship with other administrative degree programs (business and public administration) and academic disciplines (economics) where there exists additional expertise regarding the skills necessary for efficient management of caseloads and organizations.

As a result of these two shifts, it is possible that the notion of social work administration as a distinct field might undergo some change. Efficiency would bring the values and actual practice of social work into sharper focus. The practice of social work and social work administration could be helped immeasurably when carried out according to efficiency criteria. Acceptance of efficiency is the necessary first step. Educating future social workers and administrators is more critical for as in any policy, nothing happens until implementation occurs.

REFERENCES

Barnard, C. (1938). *The functions of the executive*. Cambridge, MA: Harvard University Press.

Blostein, S. (1985). A day late and a dollar short. A national survey of the testing of financial management in schools of social work. *Journal of Social Work Education*, *21*(2), 34-42.

Council on Social Work Education. (1987). *Summary information on MSW programs*. Washington, DC: CSWE.

Faherty, V. (1987). The battle of the Ms: the MBVA, MPA, MPH, and MSW. *Administration in Social Work*, *11*(2), 33-43.

Gummer, B. (1988). The hospice in transition: Organizational and administrative perspectives. *Administration in Social Work*, *12*(2), 31-43.

Gummer, B., & Edwards, R. (1988). The enfeebled middle: Emerging issues in education for social administration. *Administration in Social Work*, *12*(3), 13-23.

Hart, A. (1988). Training social administrators for leadership in the coming decades. *Administration in Social Work*, *12*(3), 1-11.

Hasenfeld, Y. (1983). *Human service organizations*. Englewood Cliffs, NJ: Prentice-Hall, Inc.

Hooyman, N. Fredriksen, K., & Perlmutter, B. (1988). Shanti: An alternative response to the AIDS crisis. *Administration in Social Work*, *12*(2), 17-30.

Malka, S. (1989). Managerial behavior, participation, and effectiveness in social welfare organizations. *Administration in Social Work*, *13*(2), 47-65.

Neugeboren, B. (1987). Introduction to the special issue: Legitimacy, effectiveness, and survival in macro education and practice. *Administration in Social Work*, *11*(2), 1-4.

Okun, A. (1975). *Equality and efficiency: The big tradeoff*. Washington, DC: The Brookings Institution.

Packard, T. (1989). Participation in decision making, performance, and job satisfaction in a social work bureaucracy. *Administration in Social Work*, *13*(1), 59-73.

Patti, R. (1982). Applying business management strategies in social agencies: Prospects and limitations. In S.H. Akabas & P.A. Kurzman (Eds.), *Work, workers and work organizations*. Englewood Cliffs, NJ: Prentice-Hall, Inc.

Patti, R. (1983). *Social welfare administration*. Englewood Cliffs, NJ: Prentice-Hall, Inc.

Patti, R. (1985). In search of purpose for social welfare administration. *Administration in Social Work*, *9*(3), 1-13.

Patti, R. (1987). Managing for service effectiveness in social welfare. *Administration in Social Work*, *11*(3/4), 7-23.

Pruger, R., & Miller, L. (1990). Efficiency and the social services: Part B. *Administration in Social Work*, *15*(1/2).

Rimer, E. (1989). Administrative decision making at youth guidance, final report (unpublished manuscript).

Russell, P., Lankford, M.W., & Grinnell, R.M. (1984). Administrative styles of social work supervisors in a human service agency. *Administration in Social Work*, *8*(1), 1-16.

Scurfield, R. (1980). Social work administrators: Their educational preparation, value orientation, role transition, and satisfaction. *Administration in Social Work*, *4*(2), 47-59.

Weinbach, R. (1990). *The social worker as manager*. New York: Longman.

Weiner, M. (1982). *Human services management: Analysis and applications*. Homewood, IL: Dorsey Press.

Wolk, J., Way, J., & Bleeke, M. (1982). Human service management: The art of interpersonal relationships. *Administration in Social Work*, *6*(1), 1-10.

PART 4: OTHER PERSPECTIVES AND VIEWPOINTS

Efficiency as a Function of Culture

Aaron Wildavsky, PhD

Before we can see how and why efficiency varies with culture, we must establish ground rules for intelligent consideration of this subject. The first rule is that *the objectives sought matter much more than how efficient are the actions undertaken pursuant to them*. After all, there are many objectives, e.g., torturing people at lowest cost, that we do not want done at all, let alone efficiently. Because it is often not easy to determine whether an objective is benevolent, we add a second rule. *Not only the objectives but the instruments of action must be morally desirable*. Imprisoning people to change their eating habits would be wrong no matter how efficient this instrument of policy turned out to be. Because objectives may be benevolent and instruments of policy may be desirable, but the results nevertheless counter-productive, perhaps persuading clients to stay on rather than get off welfare, a third rule of effectiveness should also apply: *knowledge of cause and effect* (the

Dr. Wildavsky is Professor of Political Science and Public Policy, University of California at Berkeley, Berkeley, CA 94720.

new treatment and the policy outcomes it likely generates) *is required*; otherwise it would be possible to call actions efficient that do great damage. Before actions may be called efficient, therefore, their objectives must be desirable, their instruments acceptable, and their efficacy demonstrable.

An efficient action is generally taken to be one that, compared to alternatives, achieves desirable objectives and at least costs or achieves as much of a desirable objective as can be had from a given resource. Were we all agreed on what constitutes desirable objectives compatible with our values, culture would not count; yet, were we agreed, it is surprising that we do not all come to feel and be alike. Then there would be only one culture, or way of life, and human behavior would long ago have become homogenous.

Suppose, however, that there is more than one way of life that is deemed socially desirable; then the adherents of these different ways might have different objectives and prefer different instruments of policy and, therefore, make different evaluations of efficacy. Consequently, *what is efficient for people who prefer one way of life may not be efficient for people who value a different culture*.

Consider the proposition that *rational people support their way of life*. Individualists, for instance, might seek a subsidy to protect their businesses, but if others followed suit their ability to live a competitive life would be diminished. Thus, our competitive individualists would have undertaken actions, seeking a subsidy that was efficient if their objective were to increase their income, at least in the short-term, whereas it would be inefficient if their objective were to support their culture. In this way, we arrive at the proposition that *the same actions may be economically efficient and culturally inefficient or vice versa*.

IMAGINARY CONVERSATIONS

Awash as we are in a sea of abstractions, we can now begin to make sense out of the innumerable unsatisfactory conversations we have had and heard on the subject of efficiency. Remember them: "Businesses are efficient, but social workers are caring"; "I don't want to be efficient; I want to help my clients." "Penny pinching never helped anyone." "However much my clients receive, it's always too little." "I can't be efficient and get to know my clients all at the same time." "How can it be efficient if there is never enough to go around?" "If I'm efficient, I'll only give more people lousy service." "To err is human; to be efficient is the opposite."

Unlocking the culture codes will enable us to understand what these aphorisms mean, i.e., who is saying "what" to "whom." Knowing about culture will not enable us to choose which conception of efficiency we prefer under which conditions. But it will tell us how many conceptions of efficiency make sense, and to whom they are sensible. First, a conception of culture.

Cultures are conceived as ways of life. To be viable, therefore, cultures must be livable. Livability-cum-viability requires that people must be able to justify their behavior to each other. These justifications are variously called ideologies, world views, cosmologies, or what, following Mary Douglas, cultural theorists call cultural biases.

The term "bias" signifies that each way of life is slanted in a certain direction. Its purpose (or function) is to legitimate a preferred set of social relationships. The conjunction of biases that justify relationships, and relationships that fit biases, we call cultures.

The motive-source of cultural theory is that what matters most to people are their ways of life. They get their preferences through their adherence to different ways of life. By building their institutions, people decide what they ought to prefer.

A basic axiom of cultural theory is that there are only five viable ways of life in which social relations can match cultural biases. It follows that each way of life has its own legitimating rationality. Because rational people defend their way of life, therefore, they also defend their own notion of efficiency. Thus, there are only five types of efficiency that fit the sense of rationality that inheres in each of the five ways of life (Thompson, Ellis, & Wildavsky, 1990).

For the purposes of understanding efficiency, the views of human and physical nature that are adopted by adherents of each way of life are most important. To adherents of hierarchy, for instance (a way of life-stressing structured inequality within a strong collective), human beings are born bad but can be made better by good institutions. In a corresponding manner, the concept of nature in hierarchy may be described as perverse/tolerant, which is to say that if people follow the experts sanctioned by the system, mother nature will respond kindly to their ministrations; but, if they break the rules by listening to the views of unlicensed quacks, nature will exact a terrible price.

To competitive individualists, nature is (and must be) cornucopian, her abundance limited only by human ingenuity. People do not deplete natural resources; their creativity is what makes a blob of material into a resource. Believing that the good life requires substituting self-regulation for authority, individualists conceive of human nature as malleable, good when restrained by proper incentives, bad when given power over others.

Starting with the cultures of hierarchy and individualism, we can figure out which concept of efficiency would justify them. Hierarchists, as the old formula had it, believe in the triumph of process over purpose. To them, who has the right to do what is as (or more) important than what is done. When your way of life emphasizes maintaining the distances between people, maintaining the steps in the process is the purpose. Efficiency, then, to a hierarchist is accuracy, even punctiliousness, in following the rules. Don't laugh; there can be something to this if experts are learned in a body of knowledge that is in touch with the reality it seeks to cover and if the process is reasonably attuned to the purpose. Note that humans can claim whatever they want to be efficient but that does not make it so. Nature need not comply with our importunings. Advocates of the command economy without prices can say what they want, for instance, but we know now that it is an inefficient method of securing economic growth and an irrational (i.e., counter-productive) method of sustaining single-party rule.

The individualist concept of efficiency is well-known because it is equivalent to economic efficiency in the small (getting the most out of given resources or achieving a given objective at least cost) and the large (making resources grow). Generally, individualists have a substantive notion of rationally, variously expressed as "the bottom line" or being "result oriented." It's not how you do it, individualists claim, but what you accomplish, that matters. But, they, too, care about procedures, e.g., bidding and bargaining among consenting adults.

The egalitarian way of life is about diminishing differences in power among people; what women's rights, civil rights, gay rights, children's rights, animal rights, etc. have in common is the desire to reduce power differences among men and women, black and white, straight and gay, parents and children, animals and people, etc. Egalitarians conceive of human beings as born good but corrupted by evil institutions. Like what? Like the keeping-people-in-their-place hierarchies and the inequality-producing market-oriented individualists. If people are born good, nature comes pristine into the world until it is assaulted by brutish capitalist technology. Nature to egalitarians, therefore, as the latest environmentalist flyer will testify, is fragile. The least little upset, and mother nature will wreak a terrible vengeance.

Efficient action (recall that, except to social workers, efficiency is commonly believed to be good for whatever ails you) to those whose primary value is greater equality of condition is *egalitarian action: that action is best that reduces differences among people.* Since the clients social workers serve are usually (though not always) worse off than they are, their tendency toward egalitarianism leads them to wish to take more from tax-payers so as to give more to their poor clients. When social reformers say that problems of homelessness or drug addiction or personal disorientation can only be solved or mitigated by alleviating the poverty in which these problems are spawned, they are saying, in effect, that inequality is the source of evil. Thus, egalitarians, like individualists, are result-oriented, only the results they favor differ. Where individualists are largely concerned with producing new wealth, by contrast, egalitarians are largely concerned with redistribution of existing wealth. Action efficient for maintaining social differences, as in hierarchy, might well be inefficient for increasing differences in achievement, as in individualism, or reducing differences as under egalitarianism.

Fatalists are people who believe that they can do nothing to affect their fate. To them nature is random. There is no way of discerning its pattern, hence no way of intervening with positive results. They could do better in life, perhaps inherit a legacy, but not by their own efforts. As fatalists see it, human nature is also unpredictable. The only certainty is that other people will look after themselves, not you. Mutually supportive relationships are out of the question.

One need not look far for the fatalist concept of efficiency; either it does not exist because, the world being random, no efficient action is possible, or it invokes apathy; since no action avails, inaction is best. Efficient action is just enough action to get out of the way.

Last, but by no means least instructive are the adherents of the autonomous culture, peopled by hermits, who opt out of social life since they desire neither to coerce nor to be coerced, neither to manipulate nor be manipulated. What has to be understood about hermits is that their ideas depend on not having to do anything about them. First of all, their ability to be what they are depends on the rest of us being what we are. There are three active cultures (egalitarianism, hierarchy, and individualism) and two inactive cultures, the fatalists who withdraw from within (essentially being controlled by the other ways of life) and the autonomous who withdraw outside of social involvement. Not willing to choose, they see virtue in all ways of life, human nature as plastic, and physical nature as "all of the above."

Evidently, just as there is more than one type of rational behavior, there is more than one kind of efficiency. According to their preferences for different ways of life, people may conceive of efficiency as following correct procedures and experts (the hierarchical way, applying human ingenuity to the growth of resources (the individualist way), diminishing differences among people (the egalitarian way), or avoidance of danger (the fatalist way). Getting "out of harm's way" is certainly different than doing no harm (the autonomous way).

Now we have answered our first question – efficiency for what? By answering "for justifying each way of life" a criterion of coherence can enter here: Is the user of the term "efficiency" sufficiently aware of its purpose so as to give the appropriate meaning to shore up a given way of life? That there is no single correct usage for all purposes does not mean there are not different correct usages, given the different way of life one has in mind.

Because the correct use of "efficiency" is relative to the way of life one prefers, *each concept of efficiency is good only for some purposes, not for all purposes*. Bringing down test scores of the highest achievers may be an efficient way of achieving greater equality of outcomes, possibly, though not necessarily, as efficient as raising scores of the lowest level of achievers. Neither mode of achieving efficiency, however, would necessarily be efficient for increasing the proportion of high or low achievers. In addition to a criterion of correspondence, therefore, the uses of efficiency are subject to an ethic of consequences; whatever values you seek to achieve, being efficient will get you more and being inefficient will get you less of them. *If you want to keep your promises, therefore, you need to know how to act efficiently for the purpose you have in mind.*

Why, then, to reiterate the questions with which we began, do people differ about the meaning of efficiency? Because, I answer, the cultural theory tells us they have different purposes in mind. Why do we differ so much about the desirability of being efficient. Because we do not wish to achieve purposes that are not ours. Being efficient in support of a way of life to which we do not adhere does not recommend itself to us. When we do believe in the way of life for which we are trying to be efficient, being efficient signifies being a moral actor. It is conflicts about consequences: Do we want equality so as to share out misery? Do we want growth if we also grow apart? Do we want to follow the experts when they have learned incapacities?

All of us can see ourselves in the adages about efficiency current in our society. In this respect, social workers, as people whose main business it is to understand social relations should know that ambivalence about efficiency is normal. (If "what's wrong with people" is the question, "other people" is usually the answer.) Satisfied that your work is worthwhile, being efficient in doing it makes sense.

REFERENCE

Thompson, M., Ellis, R., & Wildavsky, A. (1990). *Cultural theory*. Boulder, CO: Westview Press.

Efficiency in Social Service Administration: Historical Reflections

James Leiby, PhD

The concept of efficiency seems obvious and the criterion obviously good (who would justify a practice or program as "wasteful"?), but it is a puzzling irony that narratives of efforts to apply the concept and the criterion are often cycles of illusion and disillusion. Skeptics or opponents do not challenge the goal but argue that what the proponents want to do is not really efficient. In historical perspective, the significant observation is that sets of proponents and opponents have thought and acted in quite different historical contexts. It may be helpful to reflect on analogies between our present hopes and doubts and earlier efforts. In fact, our current definition of "efficiency" did not emerge until the 1850s. It referred then to physics and mechanics – to the ratio of fuel-input and work-output of a steam engine, for example. Later, it was generalized to industrial and agricultural technology, then to "business," and then to various bureaucracies that seemed to have much in common with business. The concept has been obvious for less than a century, and there is reason to believe that our understanding of its many applications is still in an early stage, especially with regard to social service administration and social welfare.

The noun "efficiency" comes from the adjective "efficient"; their root is the Latin verb *facere*, to make or do. Before 1700, "efficient" ordinarily modified the word "cause" to label the power that actually brought about a result, as distinct from other

---

Dr. Leiby is Professor Emeritus, School of Social Welfare, University of California at Berkeley, Berkeley, CA 94720.

factors that were present but not sufficient. This meaning occurred especially in Aristotle's distinction between four types or orders of cause (material, formal, efficient, and final) and in cosmological/theological discussions of determinism and predestination (Collingwood, 1960).

Our contemporary use had antecedents in the thinking of 18th century scientists (curious gentlemen, not academic professors) about the work of mechanics and machines, engineers and engines, which economic entrepreneurs were introducing into the market economy. There were many entrepreneurs in France and England and much excitement about their achievements and prospects. In 1885, the historian Arnold Toynbee would sum up events in the memorable phrase, "the industrial revolution." He referred to the substitution of power-driven machinery for handicraft operations. Power-driven machinery meant especially the steam engine, in place of horse power, ox power, wind and water power, and human brawn. In this and related contexts emerged the definition of efficiency as "a ratio of useful work performed to the total energy or heat taken in" (O.E.D., 1933).

The hope for science in industry fostered the Institutes of Technology and the Agricultural and Mechanical Colleges (A and M's) that appeared in the 19th century, and also the incorporation of scientific studies into conventional universities. In the 1880s, Frederick Winslow Taylor, a mechanical engineer, applied the idea to the management of a workplace or industrial plant, including functions, machines, and workers, and went on to formulate principles of "scientific management," as he called it. In practice, this meant a very thorough and detailed analysis (a "time-motion study," for example) of operations that had been shaped largely by happenstance and "rule-of-thumb" (Copley, 1923).

Meanwhile, businessmen who managed very large plants or several plants or businesses were thinking about their work. By 1920, they sometimes labeled their function "business administration" to distinguish it from the engineering side of plant management. They defined a range of concerns that went beyond technology, to finance (balancing the books, cash flow, raising capital), insurance, marketing, personnel (recruiting and holding a skilled and disciplined work force), and economic policy (getting government to work with them and not against them). Their main tool was accounting, which they borrowed from merchants and bankers in large-scale national and international markets. Their indicator of efficiency was profit and loss (the "bottom line"), and they tried to understand how all the parts of their business affected it. They developed a special kind of "cost accounting" to help them decide and plan (R. Brown, 1905).

There was an affinity between the engineer's notion of efficiency in an industrial process and the businessman's notion of efficiency affecting the return on investment. Engineers and businessmen played a competitive game; thinking about efficiency in a large way suggested strategies and tactics that gave them an advantage. The general pattern of thought might also apply to government, medical care, education, and "charities and correction," which were not subject to market competition in the ordinary sense. Many leaders in those institutions were already thinking along "scientific" lines about clearer goals and better methods. But, before considering public administration and social administration, it is worthwhile to notice that even among engineers and businessmen the new thinking encountered much resistance; moreover, employees who were not engineers or accountants were often outraged, and many deep thinkers about religion and politics shared their outrage.

Why should engineers and businessmen be skeptical about a science of management that would institutionalize, as it were, the idea of efficiency? In 1700, the typical mode of industrial production was a handicraft system. Its leaders were master craftsmen, whose interest in business was incidental to their work. The lore of their craft (sometimes called a "mystery") was typically secured in a social organization that was part of a subsistence rather than a market economy and managed labor by a system of apprenticeship. The system had ideals and satisfactions that are like those of "professions" in the 20th century. It usually changed as entrepreneurs saw opportunities for a large-scale, specialized production for a market, which usually began with some local advantage in agriculture or mining. As a rule, the earliest ideal of a production manager was someone who knew the tasks and tools because he had learned by doing. Blacksmiths, weavers, or gunsmiths combined managerial functions around experiential learning about production (Urwick,

1949; Pollard, 1965). Their ideas about efficiency (which did not go by that name) focused on incremental change in a process they knew and understood very well. In this way, a blacksmith might improve a forge, a weaver a loom, or a gunsmith a blunderbuss.

The matrix of modern thinking about productive efficiency was large-scale, market-oriented shops. The matrix of thinking about business management was large-scale agriculture or mining under the eye of a landowner, often an aristocrat who was lord over many manors. The landlords were uneasily allied with merchants who marketed their products at a distance and loaned them money for consumption, investment, or dynastic and military affairs. In the 16th century Italian merchants and bankers had begun to use double-entry bookkeeping, which spread into Northern Europe. It was a useful refinement, but well into the 19th century most business accounts were simple records of transactions intended to keep owners and managers aware of their solvency. Most businessmen lacked the inclination and patience for fancy accounting, as most breadwinners lacked the inclination and patience for personal financial planning (Urwick, 1949; Pollard, 1965).

What did get attention was the hired hands in the mills, factories, and shops that were work-sites in the market economy. In the abstract long view, power-driven machinery was much more productive than handicrafts, but in the concrete short term, its "efficiency" involved a new discipline for workers: the machine process suggested demands on the mill workers which put owners at odds with the "operatives," as they were called. Machine technology displaced craftsmen or artisans, whose anxiety stimulated a long series of "labor reform" movements in the 19th century. Many writers and artists imagined workers in the machine technology as servants or slaves of mechanical monsters. Many thoughtful people went beyond that story to see the workers as servants or slaves of the owners of the machinery (Leiby, 1960).

In fact, the efficiency of the labor force was an important concern. Frederick W. Taylor began his career with a paper for a society of mechanical engineers on a plan to pay by piece work as an incentive for more efficient work. His contribution was to analyze and measure the tasks very carefully (the origin of the "time-motion study") so as to define in an objective way what a reasonable output was, then pay workers according to that standard – higher if they exceeded it, as they might well do. Taylor had practical shop experience at all levels, and he thought of himself as rewarding the energetic, efficient worker, but piece work and the time-motion study irritated most workers. When labor unions appeared, they opposed it (Nadworny, 1955).

Taylor at the time managed a steel mill in Pennsylvania, and a wealthy ironmaster of that state, Joseph Wharton, endowed the first college-level "School of Commerce and Finance," at the University of Pennsylvania (1881). The university faculty was sniffy about this venture (Sass, 1982). The institutes of technology and mechanical engineering (like Stevens Institute in Hoboken, New Jersey, where Taylor got his degree) offered many "shop courses" of the sort that in a few decades were introduced into high schools, much as proprietary (money-making) "commercial colleges" or "business schools" offered courses in business English, business law, typing, filing, and bookkeeping, that later became a high school curriculum. All this was below the dignity of the College or University. (Academic High Schools of the late 19th century were ordinarily elite college preparatory schools that taught Latin, geometry, etcetera.)

In short, when many engineers and businessmen rallied to the cause of "scientific management," as they did after 1900, their notion of sciences was the kind of rationality manifest in a "time-motion study," bookkeeping and accounting, business law, insurance, and some economics, especially about banking and finance. The concept of efficiency was applied to operations that were, or seemed, concrete and rudimentary.

Efficiency in industrial production and business management became a conscious goal of administration in the context of a large-scale bureaucracy, in which shop management, accounting, marketing, finance, and investment were to some degree separated from "production." It is remarkable that two other large bureaucracies of the 19th century did not advance a scientific notion of efficiency: the military and the Church. It is true that West Point's military engineers, trained to build roads and bridges, often enlisted as civil engineers to build the railroads, and in 1903, after the Spanish-American War, the Army established a general staff and a War

College that might ponder administrative efficiency (Van Riper, 1987). It is true that some Church leaders thought about large-scale administration, especially in charities and missions; they thought about propaganda (in the original sense of the word) and proselytizing. But, in general, the military and the Church hoped to train charismatic leaders who would build on collective zeal rather than on a scientific approach to bureaucratic efficiency.

Civil government played a more constructive role. In Europe, the civil service, like the military service and the established Church, recruited leaders from university graduates, children of the wealthy, who looked for some sort of professional career. But, in the United States, after the rise of political parties in the "Jacksonian" period, government jobs and contracts were the "spoils" for which rival factions of politicians competed. The winning party rewarded its workers and supporters with jobs, contracts, and favorable treatment of taxes and services. Those who benefited supported the party leadership (government employees kick back part of their salary, for example). The business of American civil government got done, but the spoils system made the cost high and the quality low. It travestied democracy and outraged good citizens (Van Riper, 1987; Waldo, 1948).

With regard to bureaucratic efficiency ("good government"), the reform movements of the late 19th century focused on personnel ("civil service reform") and financial accountability (bookkeeping and auditing). The reforms took hold slowly, as party competition forced party leaders to act in spite of themselves, mostly in the Federal Government (Cleveland, 1909, 1913).

The government bureaucracies that most interest us were called, in the 19th century, "charity and correction." Those that had a self-conscious interest in administrative efficiency were mostly sponsored by state government. They were usually created at the urging of philanthropic reformers (like Dorothea Dix). In theory, they served a particular type of case by a rational program; this distinguished them from the catchall custodial jails and almshouses of local government. They were associated with a tradition of religious service; many of them were modeled on "private" or "voluntary" institutions that were separate from the political spoils system. These state institutions were protected, in theory, from the worst excesses of the spoils system by an institution of "honorary" (unpaid) boards of managers, like the boards of wealthy philanthropists who oversaw private charity. The boards of managers were expected to pick an institutional executive on the basis of technical ability and to carry into the public agency some of the spirit of service that was supposed to characterize private agencies (Leiby, 1978).

The kinds of people who showed up on the boards of state institutions were much more numerous in the galaxy of private charities that took form in the larger cities. Some of them gathered in the "charity organization societies" that, in the 1880s, began to introduce more system and businesslike methods into "voluntary" charitable work. It was a scandal, they thought, that the conscientious efforts of conscientious people should be wasted or abused (the abuses of beggars and of phony charitable promoters received much attention).

The charity organizers envisioned an ideal of scientific charity that inspired professional social work. The point to emphasize here is that they were a small fraction of a small fraction of the community. They were often officious critics of the complacent futility that characterized much charitable administration. They looked upon charitable helping as more than a means to save souls or otherwise bolster and justify sectarian religion. They criticized a kind of spoils system in private and religious charities, which might be a toy of weighty contributors or parishioners. Leaders of the traditional churches and charities, on their part, emphasized the sentiments of repentance, forgiveness, and love that were suppose to issue in charity, and they were likely to distrust considerations of scientific efficiency in its practice.

In 1865, many people interested in state charities helped found the American Social Science Association and organized a "section" of its annual meetings. In 1874, they separated themselves into the National Conference of Charities and Correction; and in 1881 advocates of charity organization began to take part in the meetings. The National Conference became the foremost of many more specialized conferences that were interested in the administration of social services. (The American Social Science Association was gradually pushed aside by associations of academic professors as they came into being.) In 1898, the Charity Organization Society of New York established a program of instruction (a professional school) for charity workers; philanthropists in other cities (Baltimore, Philadelphia, Boston, St. Louis, Chicago) soon followed suit.

After 1900, the groups of people interested in applied science and (what interests us) better social programs and administration grew and merged, focused and celebrated by magazines that found readers in the business and professional class. They espoused "efficiency" in ways that historians, looking back on the period, have labeled "cult" and "craze" (Haber, 1964; Hays, 1957; Callahan, 1962).

The most prominent form of the enthusiasm was for "economy and efficiency" in government, which was linked to academic "political science" and a profession of public administration. The excitement began with a venerable charity in New York, the Association for Improving the Condition of the Poor. In 1901, its board appointed William H. Allen as executive. He had a Ph.D. in economics. (The supply of economists was larger than the demand in those days, and many went into charitable administration, like Amos Warner in Washington, DC, and Edward T. Devine, who directed the New York COS from 1896 to 1917.) Allen became Secretary of the New Jersey State Charities Aid Association, a group of philanthropists who wanted to reform public institutions. When he moved to New York, he persuaded his board chairman, R. Fulton Cutting, a leading banker, to fund a Bureau of City Betterment to study municipal services. He soon made headlines by arguing that his agency could run a municipal bathhouse much cheaper and better—more efficiently—than the political spoilsmen who were in charge. Cutting then enlisted his friends Rockefeller, Carnegie, and Harriman to fund a Bureau of Municipal Research (1907). The Bureau became one part of the Progressive Movement by contracting to study municipal or state governments or services and propose changes (reforms) on scientific principles (Dahlberg, 1966).

The emphasis on municipal governments reflects a situation that today seems paradoxical: from a political point of view—with regard to jobs and contracts—big-city government raised and spent more money than state government. (The budget of the City of New York in 1907 was three or four times the budget of the State government.) Moreover, municipal services were like private business in many ways. There was a lively argument whether local "utilities"—horse cars, water, gas—should be public or private.

The movement reached a high point in 1911 when President Taft appointed an Economy and Efficiency Commission to study the Federal government (Taft had enhanced his reputation by his efficient administration of the Philippines). There soon followed similar commissions in New Jersey, New York, Illinois, and many other states, most staffed by veterans of the New York Bureau (Weber, 1919). The Bureau began a formal training program in 1911; by 1922, when it took the name National Institute of Public Administration, there were 25 such agencies; by 1930 there were 62 (Gulick, 1928; White, 1933).

The theoretical base of the economy and efficiency movement, as laid down by Frank Goodnow, a political scientist at Columbia, in his influential book, *Politics and Administration* (1900), was that the function of deciding on policy and the function of executing policy ought to be separate. Policy decision was properly the business of politicians, who appealed to the wisdom and preferences of the electorate. But to implement policy was a technical problem best handled by a technically competent team. His argument built on the older argument for a non-partisan civil service (Goodnow, 1900). It rested on the obvious and embarrassing fact that government administration—particularly city government—was much better in Europe than in the United States. The spoils system was a failure, from this point of view, because it combined policy and administration: administration was subordinated to the interests of party politicians, who used its jobs and contracts to sustain and advance the interests of the party and of themselves as party leaders.

Goodnow—and later partisans of economy and efficiency and scientific public administration—added new objectives to the older goal of throwing the rascals out. They were aware of the technical side of "scientific management" and more useful methods of accounting (and accountability) in business corporations. They were also aware of the increasing demand for more and better government functions and services. Their hope was not simply to prevent the wrongdoing of the spoils system, but to improve both policy and administration by clarifying the functions and goals and giving policymakers (ultimately the electorate) and administrative executives an improved capability to do the job.

Allen shows the spirit of the movement in his book *Efficient Democracy*, published in 1907 as the tide was rising. He addresses the public at large, urging readers "To apply efficiency tests to [themselves] as governor and governed." His target is not simply spoilsmen and corruption, but what he called "the goodness fallacy"—the belief that "good men will administer well"—and, therefore, the problem of administration (and of social progress) is to elect or appoint good people. By "good" he means *moral*—upright, conscientious, and honorable. He was reacting, I think, to a cloying sentimental moralizing that was conventional in the late 19th century, especially in charity work, in which good ladies gushed about the good work that good ladies were doing, and also to a moralizing politics that set heroes against villains but neglected to consider the technical side of problems and solutions. "The average citizen honestly in favor of good government does not understand," he says, "that there are an intelligence and an efficiency as far beyond the reach of mere goodness as is business efficiency beyond the reach of mere good intention." The "efficiency test," by contrast, requires careful inquiry, measurement, and comparison. His "most important suggestion," an Institute for Municipal Research, will be a "mechanism for learning and publishing the facts of social life and public administration" (Allen, 1907, 1912). The book is a series of chapters on "efficiency tests" in medical practice, hospitals, schools, charities, crime prevention, religion, and philanthropic bequests, as well as government. (His chapter on charities is mostly a review of his administration of the Association for Improving the Condition of the Poor—better recording, performance measures, and fund-raising.)

With regard to bookkeeping and accounts, the economy and efficiency commissions promoted a unified budget prepared by the executive instead of the traditional system of legislative appropriations and "ways and means." Their ideal was the potent business executive looking over the big picture and eliminating overlap or duplication; increasingly they also realized the possibilities for a more rational definition of and decision among priorities. A landmark was the federal Bureau of the Budget (1921), managed by General Charles G. Dawes, a banker who had been General Pershing's finance officer in the American Expeditionary Forces in 1917 (Dawes took the job in preference to Secretary of the Treasury, because he believed that, while nominally subordinate, it would in fact be much more important.) (Willoughby, 1918; Price, 1983).

The War had stimulated most forms of bureaucratic social organization in the United States and much thinking about them. In the 1920s, the cult and craze of efficiency evaporated. I am unaware of any general historical interpretation of administration in business, government, and the social services. Particular accounts suggest that there was much consolidation (good textbooks appeared in the several fields, for example) (White, 1926; Metcalf, 1926; Warner, 1929). In public administration, it became obvious that the improvements in accounts and budgets, administrative organization (mostly centralization of powers and functions), and personnel (civil service) did not have dramatic effects. There was so much to consolidate in 48 state governments, over 3000 county governments, and city governments of every size and description, all of which seemed distinctive to their leaders. Administrative reorganization was a slow process, full of puzzles, surprises, and disappointments, requiring patience, organizational skills that were personal as well as cognitive, and luck (particular histories are full of odd contingencies).

With regard to state welfare administration, there were in the 1920s in many states notable improvements in rationalization: better division of labor among institutions; new and specialized institutions offering more resources and choice for care and treatment; much better housekeeping arrangements and safety; better qualified and organized personnel, and a better use of the work force of convicts and patients in a "state use" system; better programs of education, treatment, or care (Leiby, 1967). And yet, these improvements were usually so incremental as to bore people who were not planning, implementing, or studying them; they were the raw materials of some administrative report or academic thesis that was usually dull if not incompetent. A problem that did get partly defined or solved usually suggested new problems to define and solve.

To venture an interpretation, the incremental progress of administrative efficiency in the 1920s was overshadowed by more reflective considerations about scientific theory, on one hand, and policy on the other. This was true even in business, where technological progress in production – dramatic as it was in the automobile, electrical and entertainment industries, and agriculture, for example – was rivaled by scientific attention to the "psychological" side of personnel work, of managerial decision-making, and of advertising and public relations to promote corporate prosperity (Crick, 1959; Hale, 1980; Barita, 1960; Simon, 1947). In the social sciences there was much empirical research, quantitative and qualitative (studies by political scientists of voting behavior, for example, and by sociologists of group participation), which raised doubts and suggested questions about all sorts of human relationships. Three ambitious reforms of the Progressive years – Prohibition, woman suffrage, and immigration restriction – proved to be disappointing if not mistaken.

As for social service administration, the impulse of scientific charity was away from mere relief or custody to prevention and rehabilitation, but that kind of thinking involved theories of cause and effect that were various and dubious (heredity or environment? personal response or social influence? laissez-faire or socialism?) and which pointed beyond charities and correction to public education and public health, if not the economy or the policy. The part of charity organization that fostered the coordination and integration of agencies and their funding and public support, which found expression in the 1920s in the "community chests" for funding local private agencies and in efforts to centralize or consolidate public agencies at the local and state level, was confused by the persistent reservations about cooperation among municipal, county, and state agencies, and among private agencies, between those that were sectarian and non-sectarian (the latter were usually the more professional) (Leiby, 1978, 1967). In the 1920s, the most favorable climate for professional thought about these problems was probably the University of Chicago School of Social Service Administration. However, its leaders were preoccupied by legal-constitutional issues in public welfare administration and by improving local poor relief with a sort of case work that advanced thinkers in New York, Boston, and Philadelphia thought was out-of-date. Meanwhile, its faculty were quite separate from the momentous scholarship in that University's departments of political science, sociology, and economics (Costin, 1983; Breckinridge, 1927; Abbott, 1940).

The big picture in the social history of the 1920s was uneven multifarious change in organizations and theories, a bewildering flux of institutions and ideas. The change and tumult spread and increased during the decades of the Depression, the War, the Cold War and the Affluent Society. It is easy to project tendentious historical simplifications on this disorder; it is difficult to realize that the SNAFU the GI's endured was in large measure the experience of everyone everywhere in every decade (who imagined 1950 in 1940? or 1960 in 1950?).

With regard to economic and political history, the most significant change was probably the effort of federal officials to direct and regulate the national economy in depression, war, and post-war international growth; this was associated with extraordinary growth in economic markets and agencies involved in them. With regard to social service administration, the federal government established its own agencies for the military and veterans and the ever-growing Social Security Administration, and it influenced state and local and private agencies through a system of grants-in-aid ("cooperative Federalism"). Federal money stimulated fund-raising and spending at other levels of government and in the private sector. The result was ever-increasing administrative complexity and significant changes in thinking about administrative efficiency.

The enthusiasm engendered by "scientific management," the "economy and efficiency commissions," and scientific notions of public and business administration—all took for granted what seems in retrospect to be a rudimentary order of distinctive functions. (It did not seem so at the time, any more than cost accounting seems rudimentary to a carpenter who sets up as a building contractor and then becomes a developer.) Simplicity meant that one could expect that problems and alternatives were readily definable and somebody (an executive) could make a decision. Complexity meant that there appeared to be several ways to define one's problems and postulate alternative solutions, and several parties to a decision involved in a complicated give-and-take process. The difficulty surfaced in business administration around executive decision-making that involved a collective or group decision-making process—a committee of experts or specialists. Complexity also surfaced in planning, especially the very large-scale planning for the war effort, or, after the war, in policies of "economic development" (Henry, 1987; B. Brown, 1986).

Complexity obscured the distinction between policy and administration. It brought into view a continuum between policy decisions and their implementation and brought out the consideration of implementation in the analysis of policy alternatives. It raised complicated questions about social processes that caught the imagination of political scientists and sociologists; it raised questions about allocations and "trade-offs" that caught the imagination of economists, especially those interested in econometrics (Dahl & Lindblom, 1953). Economists literally oriented their discipline around "efficiency," following the 1935 dicta of Lionel Robbins' *The Nature and Significance of Economic Science*. Along with these interests came operations research (originally related to wartime military planning), input-output analysis, systems analysis, risk analysis, decision theory, and information theory, and other efforts to clarify complexity. These intimations of more efficiency in complex circumstances were around or plainly foreshadowed before 1960, but events since then fell into another cycle of illusion and disillusion.

The "cycle of illusion and disillusion" I refer to is a contrast often drawn between the confidence, optimism, and idealism that fostered the social legislation (and other policies) of the 1960s, on one hand, and the disappointment, doubt, and frustration with government and social legislation manifest in the 1980s. I think this contrast is misleading, but in any case it alludes to moods, ideas, and events that go far beyond the idea of administrative efficiency and technical considerations of social policy and administration.

There were, however, at least two sequences of events that are germane to the subject. One was the effort to improve budgeting, especially in the federal government; the other, efforts at evaluative research. Both were intended to inform and rationalize; neither satisfied the hopes of its promoters.

Both followed from the cumulative growth, cost, and complexity of government programs, rising steadily through the 1950s to a sort of explosion in the 1960s. Social programs were only part of these: the Federal government was involved in the Cold War and then Vietnam, and all levels of government intervened in industry, commerce, transportation, housing, environment, and other directions as well as "health, education, and welfare" or "human services." Outside of government, among the leaders in business who were riding a forty-year boom, there was increasing attention to more rational ("scientific") business administration. They supported the apparatus of research, publication, (technical and journalistic), and teaching (academic and work-related) that intrigued, encouraged, and celebrated corporate bureaucracies. These were international phenomena, in which European and Asian businesses played an increasing part.

The budgetary reforms of the period were in part an application of business (economic) decision-making to government, as when Robert McNamara introduced the "planning-programming budgeting system" (PPBS) into the Defense Department in 1963 and President Johnson extended it to other departments in 1965. It was akin to the economy-and-efficiency business-in-government spirit of the progressive years, but now there were also a host of experts in public administration and a cumulation of their experience, for public executives had their own budget analysts and so, increasingly, did legislators. PPBS proved unacceptable in the Federal government. So did "management by objective," favored by President Nixon, and "zero-based budgeting," favored by President Carter (Downs & Larkey, 1986; March, 1983). These projects, and many related ventures, illustrated the point that economic rationality and political aspirations or vested interests were usually at odds—that policy and administration were inseparable (in academic circles the hope for a profession and science of public administration gave way to a focus on "policy analysis") (Wildavsky, 1985; B. Brown, 1986; Price, 1983).

With regard to social administration (as distinct from general policy decisions by legislators and top executives), the "grant-in-aid" system of public funding required more and more frequently that grant recipients specify and measure outcomes. (So did philanthropic foundations.) In bureaucratic terms, this procedure was a form of accountability, to the funders and increasingly to the supposed beneficiaries. But, in the event it proved to be difficult to define and measure outcomes, and the results, alas, were usually negative. The evaluators intended to encourage the work, at least what was worthwhile about it, but their findings often evoked yet another defense: Especially among social workers engaged in direct services, the argument arose that evaluation research did not measure the right things in the right way, that it was a surrogate for the wrong politics, a justification for the reductions in funding after 1972 (Aaron, 1978; Reid, 1987).

The disappointment in schemes for rational budget-making (which might have provided better for social welfare) and in evaluative research (which might have validated the work) helped generate the resistance and antagonism that Robert Pruger reports. But history suggests to me some basis for optimism.

If historians could present a comprehensive reconstruction of social service administration in 1929, 1959, and 1989, I think that it would bring out significant continuities in the form of a collective experience and memory among those who think in terms of technical expertise and professional responsibility. One significant historical question is about the size and character of the groups that join such a throng, about the changing market, as it were, for ideas and practices in administration. In historical perspective, concerns about scientific management, or business administration, or public administration, or "scientific charity," that were once hobbyhorses of a few, are now the careers of many. The grant-in-aid system of public funding was intended, among other purposes, to build an administrative capability at state and local levels, and it has in fact done so. The feds dismantle their supervisory apparatus (New Federalism), but professional administrators, policy analysts, planners, and researchers appear everywhere. Many of these are in rather autonomous career lines: health, public health, mental health, education, and correction, not to mention many other social work "fields of practice." Technical and professional workers in these agencies make use of diverse academic disciplines, but they can arrive at a lingua franca and indeed they often feel prepared to seek one another's jobs. Some notion of efficiency, however vague and controverted, is historically critical in all their efforts.

Moreover, the kind of thinking about efficiency that Pruger advocates is intended to help workers, not judge them. "Evaluative research," in its early forms, implied the question, "Shall we continue the program or end it?" It measured gross outcomes—effectiveness, as it were—rather than efficiency in the sense of a criterion to decide how to exercise discretion. Efficiency in that sense should appeal to those who accept a professional responsibility to exercise discretion and who do not shrink from confronting complexity in their work.

REFERENCES

Aaron, H.J. (1978). *Politics and the professors: The great society in perspective*. Washington, DC: The Brookings Institution.

Abbott, E. (1940). *Public assistance*. Chicago: University of Chicago Press.

Allen, W.H. (1912). Training men and women for public service. *Annals of the American Academy of Political and Social Science*, *41*, 307-312.

Baritz, L. (1960). *The servants of power*. Middletown, CT: Wesleyan University Press.

Barnard, C.I. (1938). *The functions of the executive*. Cambridge: Harvard University Press.

Breckinridge, S. (1927). *Public welfare administration* (rev. ed.). Chicago: University of Chicago Press.

Brown, B., & Stillman, R.J. (1986). *A search for public administration: The ideas and career of Dwight Waldo*. College Station, TX: Texas A & M University Press.

Brown, R. (Ed.). (1905). *History of accounting and accountants*. Edinburgh: T.C. and E.C. Jack.

Callahan, R.E. (1962). *Education and the cult of efficiency*. Chicago: University of Chicago Press.

Cleveland, F.A. (1909). *Chapters on municipal administration and accounting*. New York: Longmans, Green & Company.

Cleveland, F.A. (1913). *Organized democracy: An introduction to the study of American politics*. New York: Longmans, Green & Company.

Collingwood, R.G. (1960). *The idea of nature*. New York: Oxford University Press.

Copley, F.B. (1923). *Frederick Winslow Taylor: Father of scientific management*. 2 vols. New York: Harper.

Costin, L.B. (1983). *Two sisters for social justice: A biography of Grace and Edith Abbott*. Urbana: University of Illinois Press.

Crick B. (1959). *The American science of politics: Its origins and conditions*. Berkeley: University of California Press.

Dahl, R.A., & Lindblom, C.E. (1953). *Politics, economics, and welfare: Planning and politico-economic systems resolved into basic social processes*. New York: Harper & Row.

Dahlberg, J.S. (1966). *The New York bureau of municipal research, pioneer in government administration*. New York: New York University Press.

Downs, G.W., & Larkey, P.D. (1986). *The search for government efficiency: From hubris to helplessness*. Philadelphia: Temple University Press.

Goodnow, F.J. (1900). *Policy and administration*. New York: Macmillan.

Gulick, L. (1928). *The national institute of public administration*. New York: NIPA.

Haber, S. (1964). *Efficiency and uplift: Scientific management in the progressive era, 1890-1920*. Chicago: University of Chicago Press.

Hale, M. Jr. (1980). *Human science and social order: Hugo Munsterberg and the origins of applied psychology*. Philadelphia: Temple University Press.

Hayes, S.P. (1959). *Conservation and the gospel of efficiency*. Cambridge: Harvard University Press.

Henry, N. (1987). The emergence of public administration as a field of study. In R.C. Chandler (Ed.), *A centennial history of the American administrative state*. New York: Free Press.

Leiby, J. (1960). *Carroll Wright and labor reform: The origin of labor statistics*. Cambridge: Harvard University Press.

Leiby, J. (1967). *Charity and correction in New Jersey: A history of state welfare institutions*. New Brunswick: Rutgers University Press.

Leiby, J. (1978). *A history of social welfare and social work in the United States*. New York: Columbia University Press.

Metcalf, H.C. (1926). *Scientific foundations of business management*. Baltimore: Williams and Wilkins.

Nadworny, M.J. (1932). *Scientific management and the unions, 1900-1932*. Cambridge: Harvard University Press.

Oxford English Dictionary. (1933). Oxford: Clarendon Press.

Pollard, S. (1965). *The genesis of modern management: A study of the industrial revolution in Great Britain*. London: Edward Arnold.

Price, D.K. (1983). *America's unwritten constitution: Science, religion, and political responsibility*. Baton Rouge: Louisiana State University Press.

Reid, W.J. (1987). Research in social work. *Encyclopedia of Social Work*, 18th ed., Vol. 2. Silver Spring, MD: National Association of Social Workers.

Riper, P.P. (1987). The American administrative state: Wilson and the founders. In R.C. Chandler (Ed.), *A centennial history of the American administrative state*. New York: Free Press.

Sass, S. (1982). *Pragmatic imagination: A history of the Wharton School, 1881-1981*. Philadelphia: University of Pennsylvania Press.

Simon, H.A. (1947). *Administrative behavior: A study of decision-making processes in administrative organization*. New York: Macmillan.

Urwick, L., & Brech, E.F.L. (1949). *The making of scientific management*. 2 vols. London: Management Publications Trust.
Waldo, D. (1948). *The administrative state: A study of the political theory of American public administration*. New York: Ronald Press.
Warner, A.G., Queen, S.A., & Harper, E.B. (1930). *American charities and social work* (4th ed.). New York: Thomas Y. Crowell Company.
Weber, G.A. (1919). *Organized efforts for the improvement of methods of administration in the United States*. New York: D. Appleton and Company.
White, L.D. (1926). *Introduction to the study of public administration*. New York: Macmillan.
White, L.D. (1933). *Trends in public administration*. New York: McGraw-Hill Book Company.
Wildavsky, A. (1985). The once and future school of public policy. *Public Interest*, *79*, 25-41.
Willoughby, W.F. (1918). *The movement for budgetary reform in the states*. New York: D. Appleton and Company.

In and Out of the Rabbit Hole with Alice: Assessing the Consequences of Efficiency Prescriptions

Murray L. Gruber, DSW

The list of national worries grows, but social spending is pinch-penny. Our worries run to drug abuse, AIDS, and infant addiction, to homelessness, child abuse, and long-term health care for the elderly. On top of all that, the gap between haves and have-nots is greater than anything seen for over fifty years. In fact, one in five children now grows up poor. But these social needs have collided with yawning budget deficits. To cut the deficit – and checkmate the public sector – a competitive wedge has been driven between economic growth and domestic social spending. Simultaneously, there is pressure for more of virtually every kind of service.

The story of scarcity is also the story of efficiency. Here, intuition, practicality, and economics converge: for if resources are scarce, then to be efficient is to be rational; clearly, efficiency is good, inefficiency is bad. The reasoning seems unassailable. But, it turns out that the world of efficiency is a curious one where little is really as it seems to be.

Although the world of efficiency uses a no-nonsense empiricist rhetoric, formal logic far outweighs the empirical mode. To be sure, quantification abounds; but its veracity is more apparent than real. In fact, when core ideas are meticulously examined, they seem to take flight and stubbornly resist valid, reliable measurement. As

---

Dr. Gruber is Professor, School of Social Work, Loyola University of Chicago, 820 N. Michigan Avenue, Chicago, IL 60611.

for the implementation of efficiency principles, outcome tests are unnecessary; you have but to heed efficiency logic and good consequences will inevitably prevail. There are no social costs, no organizational dislocations, only an unshakable faith that there is a free lunch, after all. The very zenith of efficiency science is the econometric efficiency algorithm, but here quantification camouflages values and ends that can only be validated by other values and ends.

At least for these reasons, it is essential to unwind a tangled skein of conceptual, methodological, and organizational issues involved in efficiency. And if we can do that, then we can also test efficiency prescriptions empirically. After all, the usefulness of advice resides in usability and results, not abstract logic.

EFFICIENCY: GENERAL ISSUES

What Does Efficiency Mean?

Scratch below the surface of the term, "efficiency," and it turns out to be a very murky, ambiguous idea that is thrown about with such reckless abandon that it can mean virtually anything anyone wants it to mean: getting more work done; cost-cutting; rooting out waste; cost-effectiveness; a favorable ratio of benefits to costs; output productivity; distribution and even rationing; and perhaps more. This is not what science and rationality are made of.

If efficiency means everything, then it means nothing. Indeed, it seems as true today as it was some thirty years ago when Herbert Simon (1958) noted that efficiency ". . . is one of the most-used and almost certainly one of the most-abused, words in the dictionary of administration" (p. 490). In fact, it is so potent a political symbol that it can be wrapped around virtually any sort of idea to rally support for it.

Two Efficiencies: Productive and Distributive

To make sense of efficiency, the first necessity is terminological clarity. The production and distribution of services are obviously fundamental. Therefore, I will concentrate on these two types of efficiency. They are, in fact, the two main types of efficiency in microeconomic theory.

Productive efficiency means getting greater output for the budget, an idea that efficiency evangelists sometimes confuse with distribution. Distribution is a different though complementary idea. It refers to an unyielding question, "Who should get services when they are costly and in short supply?" Although it is most insistent in the health field, the question rears its Malthusian head virtually everywhere.

For the time being, let us put the two types of efficiency on "hold" so that we can develop an earlier allusion to the relationship between efficiency and organization. This will lay the groundwork for a fuller discussion of production and distribution later on.

Efficiency Is Not Organizationally Embedded

The simple fact is that there is great tension between "efficiency" and "organization." While it is not often recognized, the "theoretical world" of efficiency is almost completely circular and self-referential, and divorced from the "real world" of organizational behavior. Look at the empirical world of organizations and you find rich and complex forms of interaction, a honeycomb of decision-making processes, endlessly looping communication, negotiation, and political behavior that are necessary for survival and goal attainment; but look at efficiency theory and you find it emptied out of those processes (Lindblom & Cohen, 1979; Wildavsky, 1979). Efficiency begins by defining itself; then it proceeds inductively through formal, logical, abstract principles (Diesing, 1971). Whether those abstractions are mathematical or not, systems processes as described above are invisible. Efficiency principles, in short, work the other way around from the concrete, empirical world of organizations.

Efficiency tries to bend organizations to its dictates and, in doing so, it takes hierarchy and command more or less for granted; but, owing to the separation between efficiency dictates and organizational processes, efficiency prescriptions may be stymied. Also, because the prescriptions tend to be self-contained and lack robust connections to organizational behavior, they are ordinarily closed to their own non-economic costs and systemic side-effects. This helps to account for the fact that when efficiency-implementers assess the results of productivity programs, they tend to see positive effects exclusively (Buckner & Larcen, 1985; Spano & Lund, 1986; St. Clair et al., 1985; Patti & Ezell, 1988). In contrast, organizational analysts working outside the confines of efficiency theory typically find that performance measurement programs produce a wide range of dysfunctional consequences (Blau, 1955; Campbell, 1969; Covaleski & Dirsmith, 1981; Ridgeway, 1956). The significance of these divergent perceptions cannot be underestimated.

With the foregoing issues behind us, we can now move forward to try and unravel the twisted knots of productive efficiency, after which we will assay the same for distributive efficiency. While extending the analysis of concepts, we will place major emphasis on organizational implementation and effects.

ORGANIZATIONAL PRODUCTIVITY

Productivity measures the ratio of organizational output to input with the budget as the chief constraint. Although the idea of outputs is simple enough in the market, the non-market situation presents thorny problems. Also, as we will see, the constraint set typically lacks curbs on unwanted side-effects such as the deterioration of service quality.

Assumptions and Measurement

Non-market outputs obviously have no prices attached; they are often intangible, and they turn out to be very difficult to define and measure. Further, the distinction between outputs and program effects is not always recognized. This leads to a mistaken tendency to define productivity in terms of goal attainment.

If productivity measurement is to have any meaning, the first assumption is that program and organizational outputs can be conceptualized and measured with validity and reliability. Further, productivity measurement requires constant comparison of output with input and the right adjustments of throughput processes (work systems). For several reasons, most organizations find this beyond their reach. To illustrate, let us look briefly at government, then the social services, after which we will observe an additional set of problems in health services.

The Search for Public Sector Outputs

The classic approach to productivity measurement in the public sector was described nearly half a century ago (Ridley & Simon, 1943). Again, a key feature is the distinction between efficiency and effectiveness, hence the necessity to select outputs that can be related to goal attainment (Simon, Smithburg, & Thompson, 1958). In practice, the most thorough effort to measure productivity is the Federal government's annual assessment of the performance of its work force. As Hatry (1984) notes, the Federal approach uses an index, rather than a single measure of quantitative output: ". . . the sum of the efficiency, effectiveness, quality, and responsiveness with which products and services are delivered" (p. 37).

While this approach may be admirable, one is struck by the discrepancy between prescription and actual organizational implementation. The productivity measurements of federal agencies fail to provide effectiveness data; they do not appear actually to attend to the quality of outputs, and to a large degree, measurement turns out to be oriented to activities, not outputs. Reviewing productivity measurement at state and municipal levels, Hatry finds that here too, it is mostly dominated by such familiar measures as number of crimes, accidents, fires, and the like.

What Is Social Service Productivity?

In the social services, productivity measurement is perhaps even more confused. Ponder, for instance, the problems involved in defining and measuring the outputs of a hospice. In fact, like "efficiency" itself, the output idea is bandied about so casually, and it is so elusive, that outputs are often defined in terms of inputs (Dalton & Morelli, 1988).

What is output? Is it a composite measure, an index of quantity, quality and effectiveness or a single factor? Dalton and Morelli (1988) take a single factor view of output and limit it to a quantitative measure of "completed service episodes" such as adoptions made, foster children placed, and clinical episodes completed. Further, they say that because quality and goal effectiveness are difficult to measure, they should be deliberately excluded from the output idea, leading them to the conclusion that ". . . even ineffective treatments contribute to the output of the profession" (p. 91). This extraordinarily curious assertion illustrates the controversial assumptions that lie just below the surface orderliness of efficiency theory.

Pruger and Miller's position (in this volume) is different, but it also illustrates the problem of concept validity in trying to define output. To Pruger and Miller, outputs and effects are synonymous, but this merely exchanges one difficult problem of definition and measurement for another. As for the question of service quality, they dispute that efficiency trades it away. The question, of course, is an empirical one, but we will leave it in abeyance for the moment; what seems reasonably clear is the fact that the output idea lacks consensual validity.

Efficiency as Symbolic Politics

Amidst a conceptual muddle, social service organizations have been under considerable pressure to institute productivity programs, and, as in the government sector, there is a striking discrepancy between productivity theory and organizational practice. Activities such as "telephone calls made," or "interviews held," for example, are passed off as productivity measures. What's going on here?

Edelman's (1964) idea of "symbolic politics" is suggestive of some answers. In fact, "productive efficiency" in social services seems to be less substance and rationality, and more the use of a valued symbol to trigger support for a political agenda. As observed, because productivity programs involve multiple and elusive variables and their relationships, agencies are understandably reluctant to undertake such programs. Thus, the political game is lubricated by confusion—also known as "cognitive complexity" and "uncertainty." But beyond that, agencies are involved in an intricate web of political interactions that drive mystifications and symbolic transformation. These interactions involve both funders and claimant groups.

Agencies are, of course, resource-dependent, and funders give the appearance of requiring ever more demonstrations of productivity. That is because in the current climate, funders are pressured to display economic rationality to the outside world. But simultaneously, funders are cross-pressured – they also face the service demands of claimants. To defend themselves, funders hedge and create protective myths: they posture about productivity, but what they actually demand is huge amounts of input and activity measures. However, input and work volume may have little to do with productivity and even less to do with goal-attainment and service quality. Parenthetically, one observes a similar phenomenon with respect to the evaluation of program outcomes. Funders appear to require of agencies that they do evaluation studies, but everyone knows that resources for program evaluation are very scarce indeed; the requirement is mostly symbolic.

As organizational standard bearers, administrators are a beleaguered breed; for perimeter defense, they conform to funders' expectations and produce great volumes of numbers to describe staff activities. These are not actually output measures but familiar counts of activity such as the number of interviews held, home visits made, treatment sessions, telephone calls, completed assignments, and the like.

Substantial amounts of time may be required to generate the numbers, but the fog of cognitive complexity, combined with the "cover" provided by the language of efficiency, enables standard-bearers to produce a "front" or an illusion of productivity. What has been called a "numbers game" is, in fact, symbolic transformation (Taylor & Bogdan, 1980; Edelman, 1964). Ginsburg's (1984) interviews with administrators are illuminating. As one says, "[It's] all management by paper. They sent it to you. You sent it back. Whatever the reality is, they don't know" (p. 120). With rare exceptions, what administrators fail to produce is real productivity data.

Work Improvement and Side-Effects

Driven to defensive myth-making by the pressures of funders and of accrediting bodies (American Hospital Association, 1988; 1989) administrators institute performance monitoring systems to produce activity measures. They also adopt a convincing vocabulary of efficiency, and they even use the term, "productivity," as if it were synonymous with "work improvement."

In this game of symbolic politics and transformation, "performance-based" systems assume a key role. These systems target the way people work by using varying doses of support, facilitation, and control. The now popular "time management" system is the most innocuous (Nichols, 1983; Sheridan, 1988). However, others are more complex and involve stiffer monitoring and oversight. For instance, increasingly, one sees demanding work-volume standards for individuals and departments and their enforcement through MBO-like systems and occasionally, "performance contracts" (Buckner & Larcen, 1985). But, if these are useful for system defense, there is no free lunch; their cost is system rigidity, a common side-effect of performance monitoring. Indeed, every student of organization is familiar with the vast documentation of the dysfunctions generated by organizational control systems, e.g., proceduralism, goal-displacement, rigidity, etc.

Motivational-Behavioral Effects of Efficiency

In addition to structural rigidities, efficiency pressures impose other non-economic costs in the form of dysfunctional motivational-behavioral effects. Over thirty years ago, for example, Ridgeway (1956) found that if staff efficiency in an employment service is measured by "head counts" of rapid placements, staff will sacrifice quality and effectiveness. And again, a couple of decades later, in describing "entrapped administrators," Joseph Campbell (1979) warned: "The more any quantitative social indicator is used for decision making, the more subject it will be to corruption pressures, and the more apt it will be to distort and corrupt the social processes it is intended to monitor" (p. 85). More recently, Ginsburg's (1984) study showed that as administrators chase the "brownie points" of indicators, they ignore program quality, and they may also pressure staff to distort or fabricate performance figures.

Productivity in Health Care

The most direct evidence of the effects of productivity pressures on services comes from studies of prospective pricing in Medicare. Owing to soaring health care costs, instead of letting hospitals set their own fees, the Federal government's system for Medicare patients regulates hospital fees for almost 500 categories called DRG's (Diagnostic Related Groups), along side of which, there is intensive productivity measurement (D'Erasomo, 1989). In the DRG system, the most sophisticated, quantitatively-oriented model yet developed for the human services, the hospital resembles the ideal of a rational entity—a monitored production machine, capable of producing more outputs for its costs while presumably reconciling or optimizing health-care quality and access. In fact, DRG's are leading the way for the entire health care system. Several states have adopted them for administering Medicaid, and some large private health insurers have already instituted similar reimbursement systems.

Productivity Lowers Service Quality

The optimization of productivity takes the budget as its constraint; it is *assumed* that other values, such as service quality will be reconciled. While all the data are not yet in, the majority of studies suggest that the efficiency pressures of the DRG system are, in fact, leading to a deterioration in the quality of health services.

One of the most controversial aspects of DRG's is the matter of discharges. A hospital's costs increase with length of stay, so that the incentive for fast discharge has led to the impression of "quicker and sicker" discharges. Indeed, length of stay has been declining (Gutterman et al., 1988), and the evidence suggests that doctors (and social workers) have been pressured to drive medical patients out prematurely and in unstable condition (and psychiatric patients under Medicaid) (Inglehart, 1986; Lave, 1988). Even with expanded home health care, once discharged, many Medicare patients are in a "no care zone," unable to locate or pay for what they need (Dougherty, 1989). Despite charges of premature discharge, one very limited study of geriatric patients has shown no rebound increase in readmissions (Rich & Freedland, 1988). However, the largest study to date on the effects of DRG's shows that the death rate for patients discharged in unstable condition was significantly higher than those who were permitted to remain hospitalized until they had recovered (Kahn et al., 1989).

Finally, although the DRG system does seem to have achieved its goals of containing Medicare costs, we again find corrupting side-effects. For instance, even in non-proprietary settings, the physician may be under organizational pressure to engage in fudging or outright misrepresentation of the DRG category in order to bring in greater reimbursement (Morreim, 1988).

In summary, while productive efficiency may be a winsome term, it is irresponsible to ignore the empirical outcomes and side-effects of productivity programs. Despite the data, the efficiency agenda is in motion: macro-economic pressures drive it; efficiency-evangelists have a stake in it; legislators manipulate it; the public likes the rhetoric of efficiency; resource controllers try to use it as leverage; and social service agencies, in particular, seem encased in a game of symbolic manipulations.

DISTRIBUTIVE EFFICIENCY

Thanks to the penurious policy of the past decade, the shortfall between need and available services dictates stern distributional decisions. If we cannot serve everyone, the pain and suffering of some children, of some drug abusers, of some elderly, must be balanced against improving the well-being of others. When it comes to health services, the question, "Who Shall Live?" is starkly dramatic. But with supplies short, the "life-boat ethic" (Macklin, 1985) pertains—only if some are thrown overboard can others survive.

The pressures of Medicare's DRG system have already produced discernable shifts in distribution. In the absence of a severity of illness index which would provide for upward adjustments of reimbursement for the very old and very sick, the DRG system has created an incentive for hospitals to avoid such patients. While their numbers are small, they account for more than half of all hospital losses (Rettig et al., 1987; Horn & Backofen, 1987; Munoz et al., 1988).

Redistribution occurs in other ways, as well. There is also evidence that "product lines" and "markets" are altered in response to the fiscal constraints of DRG's. Hospitals appear to increase the services that are more profitable (surgical services) and, simultaneously, they attempt to move into the unregulated market (Kronenfeld, Baker, & Amidon, 1989). When this is not possible, another response to inadequate DRG reimbursement rates is to close the hospital doors. In fact, eighty-one community hospitals closed in 1988—for the second consecutive year, the highest number recorded. Of those eighty-one community hospitals, fifty were not-for-profit and included nine public hospitals serving elderly and poor persons (American Hospital Association, 1989).

Quite apart from life-saving, we are increasingly being forced to think about whose suffering to alleviate when services are squeezed. You might choose to help the most desperate, while I might choose to help the most promising. How are we to decide that in principle? Economic theory gives some answers.

Welfare Economics

Above all other questions, welfare economics has wrestled with this one: how to weigh the gains and losses to different individuals. Does Smith's benefit compensate for Jones' loss?

In the theoretical realm, Pareto Efficiency advises that solution A is better than B if all individuals in the society prefer A to B. So unlikely is this that there is a fallback principle: when not all are in agreement, if the gainers from solution A could compensate the losers so that everyone would be as well or better off than before, then we would also have a "Pareto efficient" allocation. But, logic and the real world collide. In the real world, compensation is almost never paid. Efficiency logic also leads to this type of bizarre outcome: if the alternative for homeless Harry is freezing, and neither he nor anyone else in the society objects, we might offer to house Harry in a packing crate. This allocation is Pareto-efficient because no one has been made worse off.

To escape from such theoretical sterility, welfare economics has taken a much more applied tack in recent decades in the development of cost-benefit analysis and cost-effectiveness analysis. But, it is important to note that these methods deal with aggregate benefits; who gains and who loses is essentially irrelevant. So inevitably, the Procrustean question continues to haunt us: Will Smith or Jones get services when they are in short supply? Microeconomics comes up with another principle—potential improvement (Cuyler, 1973).

Under the rule of potential improvement, each increment of benefit gained is given the same weight no matter who gets it. A dollar found by a poor man has the same value as a dollar found by a millionaire. Neither person is intrinsically more deserving than the other, so the only thing that counts is the absolute size of the gain. Embodying the same economic principle, DRG's are unweighted by a "need" or severity of illness index. True, this works against the chronically ill and very sick elderly, but remember, we cannot reduce everyone's misery.

Potential improvement works in much the same way when the benefits are social and psychological. For social work, Turem and Borne (1983) have stated "potential improvement" as follows: "an agency should concentrate on programs for those clients whom it serves most successfully. . . . If an agency cannot serve everyone, its staff must remember that resources spent on the intractable client are not available for helping those they know how to assist" (p. 208).

On Saying "No"

Metaphorically, we are still in the life-boat; with short supplies, we are faced with the question of who to throw overboard. To see how potential benefit might work, let us perform some mental experiments.

In the first situation, we must decide how to allocate services for persons diagnosed as suffering from Alzheimer's disease. Let us take some known facts. There is no definitive test for Alzheimer's, nor is there a cure. Furthermore, the prevalence of clinically diagnosed Alzheimer's is much more widespread than commonly believed, and both the rate and degree of impairment are associated with age. Overall, of those 65 and older, an estimated 10.3% have Alzheimer's disease. Among 65-74 year olds, 3.0% probably have Alzheimer's, compared with 18.7% of those 75-84, and 47.2% of those over 85 years (Evans et al., 1989).

Although over a period of years, a patient's condition does tend to worsen gradually, a reasonable quality of life can often be maintained for some years if the proper services are available. These include supports for care-givers and patients such as day-care and respite services, as well as programs of cognitive and sensory stimulation, "restorative nursing," and such social work services as "reminiscence" and activity groups. But we are thwarted by an inadequate supply of such services and cannot assist all who might potentially benefit. Whose suffering will be alleviated, and who will get simple custodial care?

With scarcity in mind, let us consider another "hypothetical" situation involving the foster child-care system. The known facts are underfunding, with children lost in the system, often losing contact with their natural parents, and many suffering multiple replacements. By focusing intensive services on families in which immediate risk to children is relatively low, we may be able to prevent foster care placement when a child has been mistreated. It would appear to be not only humane but efficient to redirect resources to these families. But, inevitably, there will be children at very great risk of harm for whom foster care is essential. There is also a new stream into the foster care system – infants exposed to drugs.

We would like to skirt the reality that helping some children through "family preservation" programs may mean hurting others who are consigned to foster care; but in searching for distributional principles, we do not have the escape-hatch of added resources. Does this change our decision to redirect resources away from foster care?

For our last experiment, let us consider the homeless mentally impaired. Although data estimates vary widely, several lines of evidence suggest a rate of psychiatric impairment of about 33% among the homeless, and a co-occurrence rate of alcohol, drug and mental problems of about 25% (Wright, 1989).

There can hardly be any doubt that "potential improvement" would systematically ignore the latter 25% in favor of singly-impaired individuals. In general, the homeless mentally ill who have problems in virtually all areas of life, and who receive hardly any mental health services at all, would be justifiably deprived in favor of more reliable, well-motivated, clients. As is, the reliable, articulate, insured client has easier access to mental health services. And, potential improvement tells us this is a wise and prudent use of scarce resources. True, in practice, one does find a variety of outreach and intervention efforts to the homeless in such places as park benches and in bus depots (Belcher & Ephross, 1989), but we must avoid using this as an easy out. We are searching for a principle, so we must remember that the more we devote resources to those with multiple impairments, the fewer resources will be available to healthier, more productive persons.

Pressured Professionals and Service Distortions

Should the physician adjust his best professional judgment of patient's needs for the sake of DRG's financial incentives? Is the social work practitioner to serve only those needs that lead to economically efficient allocations?

The solid practicality of the potential improvement criterion is more apparent than real. By now it should be clear that client selection (resource distribution) based on expectations of good outcome is a slippery slope. Inevitably, it will lead to service distortions in which preference is given to individuals with favorable prognoses, and less severe, sometimes even minor, problems. Within groups, we would avoid those who suffer multiple risks; we would avoid persons whose lives are strained by problems in many areas, those from whom little gain can be expected. And, if choices must be made between groups, we would shun the old, the poor, the unemployed, the chronically ill—all those who suffer multiple handicaps. In fact, to screen out effectively such potential users, they may be made to pay opportunity costs through more or less long queues and multiple referrals, and psychic costs through burdensome application procedures and stigmatization.

Distributional efficiency exacts other organizational side-effects or costs. Already, one sees matter-of-fact administrative expectations that social workers and other professional service-givers will become efficiency-agents in their case-by-case decisions. This leads to an inevitable, if somewhat hidden, collision between organizational requirements and sanctions and fidelity to the ethic of service, and it revisits the conflict between bureaucratic and professional orientations.

There are other more subtle side-effects. The current state of knowledge and skill in relieving suffering is limited, but the search for distributional efficiency preempts innovation; the potential improvement criterion simply assumes practice "as is" and prescribes that clients be selected who fit "tried-and true" methods. This automatically and artificially inflates judgments of effectiveness and immunizes methodological shortcomings. There is little room left for pushing intervention boundaries, say for instance, in work with the homeless. This is a recipe for professional and organizational stagnation.

We would be neglectful if we did not again highlight the "constraint problem" in optimizing productive and distributional efficiency. Whether one is dealing with mental health or Medicare, the chief constraint is the budget. Herbert Simon's epigrammatic conclusion is pertinent here: "If you allow me to determine the constraints, I don't care who sets the optimization criterion" (Katz & Kahn, 1978, p. 242). It may be appealing to argue that administrators will take account of negative side-effects as constraints on efficiency, but when the squeeze is on, the data suggest otherwise. Mostly, clients, patients, and workers pay the costs.

CONCLUSIONS

To enter the world of efficiency prescriptions is to leap down the rabbit hole into an Alice in Wonderland world where everything becomes "curiouser and curiouser." Efficiency has the surface gloss of clarity and systematic prudence, and it seems to mean almost anything one wants it to mean. Efficiency uses a facade of empiricism to hide values and consequences, but true believers nevertheless bestow on it the talismanic powers of "rationality." Productivity seems clear enough until one actually tries to measure it; then it turns into motion, activity, political incantation, and symbol—except when productivity pressures drive redistribution. Then, we find a curious sort of rationality whose logic turns age-discrimination into policy, a logic that closes the door to the neediest and those who suffer the greatest risks.

Rationality must be judged by values, goals, and empirical results, not by efficiency's closed and abstract logic alone. Distributional prescriptions and algorithms should be vigilantly scrutinized for counter-equity effects and potential systems dysfunctions. Minimally, potential improvement should be adjusted or indexed to severity, risk, and unmerited handicap so that the underserved do not become even more so. To go further, a Rawlsian position might be adopted in which unequal distributions, if they are to occur, will benefit the least, not the most advantaged.

Contemporary efficiency theory is too linear and closed; as observed, it jumps from the theoretical drawing board to its own limited outcomes without including organizational processes and side-effects. One of the tasks of scholarship is to overcome the alienation of economic prescriptions from the real world of organizational behavior. That would be a significant step forward in linking theory and practice. As for beleaguered administrators entrapped by efficiency pressures, they are hard put to buffer the negative side-effects, and they are ill-served by the curious idea—all the more remarkable for its economic context—that efficiency is cost-free. In the real world, efficiency is purchased at the costs of service quality, organizational rigidity, help to the neediest, and injury to both service-givers and the ethic of service.

Perhaps, as the saying goes, "There is no problem so big we can't avoid it." Fundamentally, "efficiency" is used as a beguiling and potent political symbol in a top-level battle to restrict "social needs" and "entitlements." That the contest is pushed down to lower, more vulnerable layers of society does not lessen its political significance. To be sure, at ground level, there is no grist for the social mill in waste, ill-conceived systems, or unresponsive services. Therefore, *ad hoc*, localistic work improvements may be highly desirable. But the productive-distributive paradigm itself is much larger, and it is fundamentally ideological. It claims "rationality," but it manipulates political symbols to undercut social benefits and social rights. Surely, rationality must be made of better stuff.

REFERENCES

American Hospital Association. (1988). Healthcare information and management systems society and the society for hospital social work directors. *Conference on productivity measurement in the social work department*. Denver, CO: August 19-20.

American Hospital Association. (1989). Social work administration. *The Newsletter for the Society of Hospital Social Work Directors*, *15*, 1-27.

Belcher, J.R., & Ephross, P.H. (1989). Toward an effective practice model for the homeless mentally ill. *Social Casework*, *70*, 425-427.

Blau, P. *Bureaucracy in modern society*. (1956). New York: Random House.

Buckner, M., & Larcen, S.W. (1985). Strategies for increasing productivity and revenues in community mental health centers. *Community Mental Health Journal*, *21*, 237-251.

Campbell, D.T. (1979). Assessing the impact of planned social change. *Implementation and Program Planning*, *2*, 67-90.

Covaleski, M.A., & Dirsmith, M.W. (1981). Budgeting in the nursing services area: Its management control, political and witchcraft uses. *Health Care Management Review*, *6*, 17-25.

Cuyler, A.J. (1973). *The economics of social policy*. London: Martin Robinson & Co.

Dalton, G.L., & Morelli, P. (1988). Casemix and caseload: Measurement of output of a social agency. *Administration in Social Work*, *12*(4), 81-93.

Diesing, P. (1971). *Patterns of discovery in the social sciences*. Chicago: Aldine.

Dougherty, C.J. (1989). Ethical perspectives on prospective payment. *Hastings Center Report*, *19*, 5-11.

Edelman, M. (1964). *Symbolic uses of politics*. Urbana, IL: University of Illinois Press.

D'Erasmo, M.D. (1989). Productivity systems and DRG reporting. *Topics in Health Care Financing*, *15*, 69-79.

Evans, D.E. et al. (1989). Prevalence of alzheimer's disease in a community population of older persons: Higher than previously reported. *Journal of the American Medical Association*, *262*, 2551-2557.

Ginsberg, P. (1984) Dysfunctional aspects of performance measurement. In C. Windle (Ed.). *Program performance measurement: Demands, technology, and dangers*. National Institute of Mental Health. Series BN No. 5, DHHS Pub. No. (ADM) 84-1357. Washington, DC: U.S. Government Printing Office, 67-72.

Gutterman, S. et al. (1988). The first 3 years of medicare prospective payment: An overview. *Health Care Financing Review*, *9*, 67-76.

Harry P. Hatry. (1984). The status of productivity measurement in the public sector: An update. In C. Windle (Ed.). *Program performance measurement: Demands, technology, and dangers*. National Institute of Mental Health. Series BN No. 5, DHHS Pub. No. (ADM) 84-1357. Washington, DC: U.S. Government Printing Office, 37-50.

Horn, S.D., & Backofen, J.E. (1989). Ethical issues in the use of a prospective payment system: The issue of a severity of illness adjustment. *Journal of Medicine and Philosophy*, *12*, 145-153.

Kahn, K.L. et al. (1989). Abstract. DRG based prospective payment system and quality of care. *Clinical Research*, *37*, 315a.

Inglehart, J.K. (1986). Early experience with prospective payments of hospitals. *New England Journal of Medicine*, *314*, 1460-1464.

Katz, D., & Kahn, R.L. (1979). *The social psychology of organizations*. New York: John Wiley & Sons.

Kronenfeld, J.J., Baker, S.L., & Amidon, R.L. (1989). An Appraisal of organizational response to fiscally constraining regulation: The Case of hospitals and DRG's. *Journal of Health and Social Behavior*, *30*, 41-55.

Lave, J.R., & Frank, R.G. (1988). Factors affecting medicaid patients' length of stay in psychiatric units. *Health Care Financing Review*, *10*, 57-66.

Lindblom, C.E., & Cohen, D.K. (1979). *Usable knowledge: Social science and social problem solving*. New Haven, CT: Yale University Press.

Macklin, R. (1985). Are we in the lifeboat yet? Allocation and rationing of medical resources. *Social Research*, *52*, 607-623.

Morreim, E.H. (1988). Cost-containment: Challenging fidelity and justice. *Hastings Center Report*, *18*, 20-25.

Munoz, E. et al. (1988). Financial risk and hospital cost for elderly patients. *Archives of Internal Medicine*, *148*, 909-912.

Nichols, A.W. (1983). Teaching time management: A gender sensitive perspective. *Administration in Social Work*, 7(1), 77-85.

Patti, R.J., & Ezell, M.E. (1988). Performance priorities and administrative practice in hospital social work departments. *Social Work in Health Care*, *13*, 73-90.

Rettig, P. et al. (1987). Medicare's prospective payment system: The expectations and the realties. *Inquiry*, *24*, 173-88.

Rich, M.W., & Freedland, K.E. (1988). Effect of DRG's on three-month readmission rate of geriatric patients with congestive heart failure. *American Journal of Public Health*, *78*, 680-682.

Ridgeway, V. (1956). Dysfunctional consequences of performance measurement. *Administrative Science Quarterly*, *1*, 240-247.

Ridley, C.E., & Simon, H.A. (1943). *Measuring municipal activities*. Chicago: The International City Manager's Association.

Sheridan, M. (1988). Time management in health care. *Social Work in Health Care*, *13*, 91-99.

Simon, H., Smithburg, D.W., & Thompson, V.A. (1971). *Public administration*. New York: Alfred A. Knopf.

Spano, R.M., & Lund, S.H. (1986). Productivity and performance. *Social Work in Health Care*, *11*, 25-39.

St. Clair, C.H., Silver, M.J., & Spivack, G. (1985). An instrument to assess staff time utilization in a community mental health center. *Community Mental Health Journal*, *11*, 371-380.

Taylor, S.J., & Bogdan, M. (1980). Defending illusions: The organization's struggle for survival. *Human Organization*, *39*, 209-218.

Turem, J. S., & Borne, K.E. (1983). Doing more with less. *Social Work*, *28*, 206-210.

Wildavsky, A. (1979). *Speaking truth to power: The art and craft of policy analysis*. Boston: Little Brown and Co.

Wright, J.D. (1989). *Address unknown: The homeless in America*. New York: Aldine de Gruyter.